SpringerBriefs in Environmental Science

More information about this series at http://www.springer.com/series/8868

Thomas R. Sinclair
Editor

Water-Conservation Traits to Increase Crop Yields in Water-deficit Environments

Case Studies

Springer

Editor
Thomas R. Sinclair
Crop and Soil Sciences Department
North Carolina State University
Raleigh, NC, USA

ISSN 2191-5547 ISSN 2191-5555 (electronic)
SpringerBriefs in Environmental Science
ISBN 978-3-319-56320-6 ISBN 978-3-319-56321-3 (eBook)
DOI 10.1007/978-3-319-56321-3

Library of Congress Control Number: 2017939631

Printed on acid-free paper

This Springer imprint is published by Springer Nature
The registered company is Springer International Publishing AG
The registered company address is: Gewerbestrasse 11, 6330 Cham, Switzerland

Preface

Water deficit is the major cause of crop yield loss in virtually all areas of the globe that are not irrigated. Until very recently, there has been little progress in improving specific physiological traits in crop plants to decrease the impact of drought on yield. Two physiological traits have been recently identified that result in early-season, limited water use so that more water is available to support later-season reproductive growth. These water-conservation traits are early, partial stomata closure with soil drying and partial stomata closure under elevated vapor pressure deficit. This book reviews the mechanistic activity of both of these traits.

The initial chapters (Chaps. 1, 2, and 3) present the physiological basis of each of the water-conservation traits. These chapters provide readers with a full introduction and description of the functioning of these traits to result in putative yield increases. The remainder of the book (Chaps. 4, 5, 6, 7, 8, 9, 10, and 11) is devoted to reviews of the research on the two water-conservation traits in individual crop species. An important part of the reviews is the progress in understanding the traits so that they can be developed in each species leading to improved, higher-yielding cultivars. Commercial cultivars are already on the market for maize and soybean resulting in yield increase under water-deficit conditions. The progress in developing the water-conservation trait presented in this book would be of interest to all that are concerned about increasing future crop yields as water deficits are anticipated to become more prevalent.

Raleigh, NC, USA | Thomas R. Sinclair

Contents

1 **Introduction** 1
Thomas R. Sinclair

2 **Early Partial Stomata Closure with Soil Drying** 5
Thomas R. Sinclair

3 **Limited-Transpiration Rate Under Elevated Atmospheric Vapor Pressure Deficit** 11
Thomas R. Sinclair

4 **Soybean** 17
Thomas R. Sinclair

5 **Peanut** 27
M. Jyostna Devi and Avat Shekoofa

6 **Chickpea** 35
Kaliamoorthy Sivasakthi, Mainassara Zaman-Allah, Murugesan Tharanya, Jana Kholová, Thiyagarajan Thirunalasundari, and Vincent Vadez

7 **Lentil** 47
Michel Edmond Ghanem, Hélène Marrou, Julie Guiguitant, and Fatima ez-zahra Kibbou

8 **Maize** 55
Avat Shekoofa and Sunita Choudhary

9 **Sorghum** 65
Sunita Choudhary and Jana Kholová

10 Pearl Millet ... 73
Murugesan Tharanya, Jana Kholová, Kaliamoorthy Sivasakthi, Thiyagarajan Thirunalasundari, and Vincent Vadez

11 Wheat ... 85
Walid Sadok

Index ... 93

Contributors

Sunita Choudhary Multi-Crop Research Center, Pioneer Hi-Bred Private Limited, Hyderabad, India

M. Jyostna Devi Crop and Soil Sciences Department, North Carolina State University, Raleigh, NC, USA

Michel Edmond Ghanem International Center for Agricultural Research in the Dry Areas (ICARDA), Crop Physiology Laboratory, North-Africa Platform, Rabat, Morocco

Julie Guiguitant International Center for Agricultural Research in the Dry Areas (ICARDA), Crop Physiology Laboratory, North-Africa Platform, Rabat, Morocco

UMR-SYTEM, Montpellier SupAgro, Montpellier, France

Jana Kholová International Crops Research Institute for the Semi-Arid Tropics (ICRISAT), Greater Hyderabad, Telangana, India

Fatima ez-zahra Kibbou International Center for Agricultural Research in the Dry Areas (ICARDA), Crop Physiology Laboratory, North-Africa Platform, Rabat, Morocco

Hélène Marrou UMR-SYTEM, Montpellier SupAgro, Montpellier, France

Walid Sadok Department of Agronomy and Plant Genetics, University of Minnesota, St. Paul, MN, USA

Avat Shekoofa Department of Plant Sciences, University of Tennessee Knoxville, Jackson, TN, USA

Thomas R. Sinclair Crop and Soil Sciences Department, North Carolina State University, Raleigh, NC, USA

Kaliamoorthy Sivasakthi Bharathidasan University, Tiruchirappalli, Tamil Nadu, India

International Crops Research Institute for the Semi-Arid Tropics (ICRISAT), Greater Hyderabad, Telangana, India

Murugesan Tharanya Bharathidasan University, Tiruchirappalli, Tamil Nadu, India

International Crops Research Institute for the Semi-Arid Tropics (ICRISAT), Greater Hyderabad, Telangana, India

Thiyagarajan Thirunalasundari Bharathidasan University, Tiruchirappalli, Tamil Nadu, India

Vincent Vadez International Crops Research Institute for the Semi-Arid Tropics (ICRISAT), Greater Hyderabad, Telangana, India

Mainassara Zaman-Allah International Center for Maize and Wheat Improvement (CIMMYT), Mount Pleasant, Harare, Zimbabwe

Chapter 1
Introduction

Thomas R. Sinclair

Water deficits are the major limitation in increasing crop yields in many regions of the world. Even those areas not commonly thought to be constrained by water availability suffer intermittent water deficits that might impact yield. Hence, a major focus of both current and future research to increase crop yields needs to be on enhancing plant yields for water-limited environments.

Various plant traits that might result in yield increases in water-limited environments have been vigorously discussed for decades. One of the most comprehensive discussions was presented by Ludlow and Muchow in 1990 in which they consider 16 plant traits as candidates for drought tolerance. However, several of these were thought to be specifically useful for plant survival, which is of little interest in annual crop production. If water availability is so low that plant survival is an issue, then yields will be extremely low even if the plants survive. Ludlow and Muchow (1990) also suggested transpiration efficiency as a possible approach to increasing crop growth under water-limited conditions. Intrinsic transpiration efficiency is leaf C assimilation divided by stomatal conductance. However, only limited progress has been made with this suggestion for several reasons (Sinclair 2012). Since both terms of the ratio share part of a common diffusion pathway, the options for increase in transpiration efficiency are limited. While some genetic variation has been found for intrinsic transpiration efficiency, the results are skewed to finding genotypes with poor efficiency rather than lines that have unusually high efficiencies. The one major case of success is with wheat in Australia where an improved intrinsic transpiration efficiency led to a wheat cultivar with a yield increase of 0.11 t ha^{-1} (Rebetzke et al. 2002).

Two of the suggestions of Ludlow and Muchow (1990) focused on the seasonal dynamics of water use. The objective of these ideas was to decrease water use early

T.R. Sinclair (✉)
Crop and Soil Sciences Department, North Carolina State University, Raleigh, NC, USA
e-mail: trsincla@ncsu.edu

T.R. Sinclair (ed.), *Water-Conservation Traits to Increase Crop Yields in Water-deficit Environments*, SpringerBriefs in Environmental Science,
DOI 10.1007/978-3-319-56321-3_1

in the season so that more water is available to sustain plant activity during the critical reproductive period when seeds are growing. One of the best examples of this conservative approach to seasonal use of water is from the studies of Richards and Passioura (1989) on xylem vessel diameter in wheat (*Triticum aestivum* L.). They discovered that smaller diameter vessels restricted water flow resulting in conserved soil water and higher yields in water-limited environments. However, no commercial cultivars resulted from their research.

Recently, two plant traits not suggested by Ludlow and Muchow (1990) to achieve conservative use of soil water have been intensively investigated. At least one of these traits is now present in commercial lines of maize and soybean. Both traits are based on partial stomatal closure under specific environmental conditions to limit transpiration rate and result in conserved soil water. One trait is partial stomatal closure at higher soil water contents than normally occurs in most plants. While all plants initiate partial stomatal closure as soil dries, the usual range for the initiation of closure is at available soil water of 0.25–0.40 of total available soil water (Sadras and Milroy 1996). Select genotypes have now been identified in a number of crops that initiate stomatal closure at much higher soil water contents than the usual range. Progress in identification of genotypic variability and understanding the basis for the unusual response is a major topic of this book.

An important consequence of the early initiation of partial stomatal closure in a drought cycle is soil water conservation resulting in an extended period in the use of the soil water. A consequence of the extended period of water use is that plants can sustain physiological activity over a longer period during water deficit, more efficiently use the available water, and may be better positioned to resume full plant development and growth if the soil is re-watered. These possibilities are discussed in each of the case study species presented in this book.

The second trait to conservative soil water is to have partial stomatal closure when atmospheric vapor pressure deficit (VPD) is greater than a threshold that when expressed has been commonly found to occur in the range of 1.8–2.4 kPa. By limiting transpiration rate above a threshold VPD, which generally occurs during midday, water loss rates would be lessened to result in soil water conservation. Such water conservation, particularly if it occurs early in the growing season, would position the crop to have more water for sustained physiological activity later in the growing season and potentially increase crop yields (Sinclair et al. 2016). It appears, however, that in many commercial cultivars of crop species, this limited-transpiration trait under elevated VPD is not common. This may result from a failure to do breeding selections directly focused on water-deficit conditions. Recent investigations, however, have now identified at least a few genotypes in nearly all crop species that express the limited-transpiration trait. The identification of these genotypes and investigations into the basis for the expression of the limited-transpiration trait is the other major topic of this book.

Since both water-conservation traits result from partial stomatal closure, they both result in lost photosynthetic activity on the days when the traits are invoked. Hence, there is a trade-off of lost photosynthetic activity during expression of the trait and benefit from extended physiological activity during dry periods. In addition, conservation of soil water may result in greater soil evaporation, especially if the top soil layers remain wetter, and the full benefit of sustained plant physiological activity will be lost. Therefore, the relative benefit of the two traits depends on temporal pattern in the rainfall and severity of the drying event. Experimental and simulation studies are presented in the book to indicate the environment conditions and locations where each of the water-conservation traits may result in increased high probability of yield increase.

This book offers a review of both water-conservation traits and their potential impact on crop yield. Since there are major differences among crop species, this book is organized to present various crop species individually. These case studies include many of the major global crop species. However, before considering each of the species, the first two chapters background the nature of each of the water-conservation traits. The goal of the book is to provide a report on the current status on the water-conservation traits by partial stomatal closure at earlier stages of soil drying and under elevated VPD. In addition, suggestions for further exploitation of these two traits are discussed.

References

Ludlow MM, Muchow RC (1990) A critical evaluation of traits for improving crop yields in water-limited environments. Adv Agron 43:107–153

Rebetzke GJ, Condon AG, Richards RA, Farquhar GD (2002) Selection for reduced carbon isotope discrimination increases aerial biomass and grain yield of rainfed bread wheat. Crop Sci 42:739–745

Richards RA, Passioura JB (1989) A breeding program to reduce the diameter of the major xylem vessel in the seminal roots of wheat and its effect on grain yield in rain-fed environments. Aust J Agric Res 40:943–950

Sadras VO, Milroy SP (1996) Soil-water thresholds for the responses of leaf expansion and gas exchange: a review. Field Crops Res 47:253–266

Sinclair TR (2012) Is transpiration efficiency a viable plant trait in breeding for crop improvement? Funct Plant Biol 39:359–365

Sinclair TR, Devi JM, Carter TE Jr (2016) Limited-transpiration trait for increased yield for water-limited soybean: from model to phenotype to genotype to cultivars. In: Yin X, Struik PC (eds) Crop systems biology. Springer Intl, Basel, pp 129–146

Chapter 2
Early Partial Stomata Closure with Soil Drying

Thomas R. Sinclair

Soil drying has long been recognized as a major limitation on plant gas exchange, mediated by early partial stomatal closure. To quantify the response of stomata to water deficit, relationships have been explored between stomatal conductance and bulk leaf water potential. Unfortunately, unique relationships could not be established, likely as a result of the ephemeral nature of water potential. Leaf water potential itself is dependent on the water balance of the leaf and hence responsive to stomatal conductance. Further, the difficulty in measuring or estimating leaf water potential makes it ill-suited as an independent variable in estimating stomatal conductance in the field (Bennett et al. 1987). An alternative approach that has proved to provide a more stable and consistent independent parameter in defining water-deficit stress for plants has been soil volumetric water content (Sadras and Milroy 1996). Hence, the soil volumetric water content at which partial stomatal closure is initiated is a critical variable in comparing the sensitivity of plants to soil drying. Those plants that initiate stomatal closure at higher soil water contents causing soil water conservation and allowing water use to be spread over more days result in sustained crop physiological activity during the ongoing development of water deficit.

2.1 Background for Stomatal Response to Soil Drying

About 75 years ago, Martin (1940) and Kramer (1944) related plant water loss to volumetric soil water content. They reported that there was little change in water loss rate until about two-thirds of the extractable soil water was removed from the soil. Ritchie (1981) formalized the relationship by suggesting that expressing water

T.R. Sinclair (✉)
Crop and Soil Sciences Department, North Carolina State University, Raleigh, NC, USA
e-mail: trsincla@ncsu.edu

T.R. Sinclair (ed.), *Water-Conservation Traits to Increase Crop Yields in Water-deficit Environments*, SpringerBriefs in Environmental Science,
DOI 10.1007/978-3-319-56321-3_2

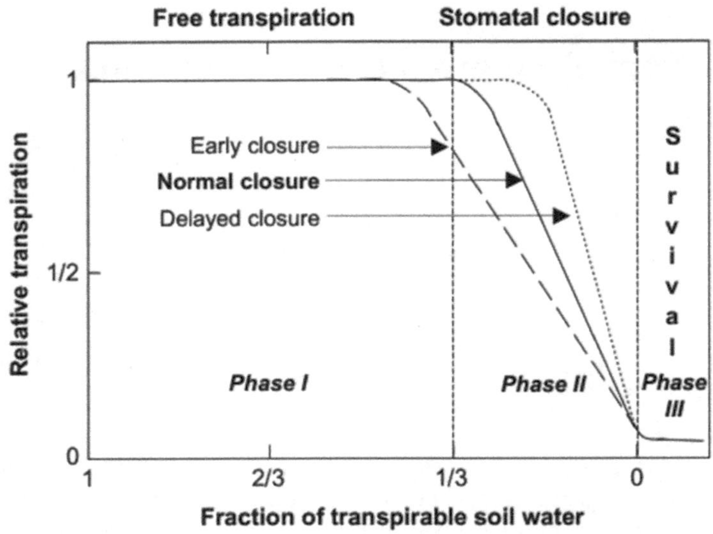

Fig. 2.1 Plot of transpiration rate of plant subjected to soil drying relative to well-watered plants versus the fraction of transpirable soil water (FTSW) remaining in the soil (Sadok and Sinclair 2011). The FTSW value of 1 is pot or field capacity, and a value of 0 is when relative transpiration has reached a value of 0.1 indicating little leaf gas exchange, and transpiration rate is no longer decreasing linearly

loss as a fraction of the extractable soil water resulted in a similar response function that was appropriate for a number of crop species and under a range of conditions.

Figure 2.1 illustrates the changes in transpiration as soil water content decreases. In this illustration, the amount of water in the soil is expressed as the fraction of the water that is used by the plant to support transpiration, i.e., fraction transpirable soil water (FTSW). Phase I is when water is readily available to the plant, and there is usually no limitation on transpiration rate. However, eventually the soil dries to a point, usually around 1/3 FTSW, where transpiration rate decreases. This point is the initiation of Phase II in soil drying. In Phase II, transpiration rate decreases approximately linearly with further decreases in soil water content until soil water available to support transpiration is exhausted, i.e., FTSW = 0 (Fig. 2.1). At FTSW = 0, the plant is substantially stressed and in nearly all cases wilted, but senescence has not yet occurred. Phase III is the survival phase where stomatal closure allows low amounts of water loss rates as defined by the leaf epidermal conductance. Of course, under this situation there is very little or no CO_2 diffusion into the leaf.

The transition from Phase I to Phase II generally occurs at about 1/3 FTSW. Sadras and Milroy (1996) concluded in their review that the range for this transition was about 0.25–0.40 fraction extractable soil water. The basis for such a general pattern in transpiration response to soil drying was suggested by Sinclair (2005) to be a consequence of decreasing soil hydraulic conductivity with soil drying. Figure 2.2 shows the calculated response in relative transpiration rate with decreasing FTSW

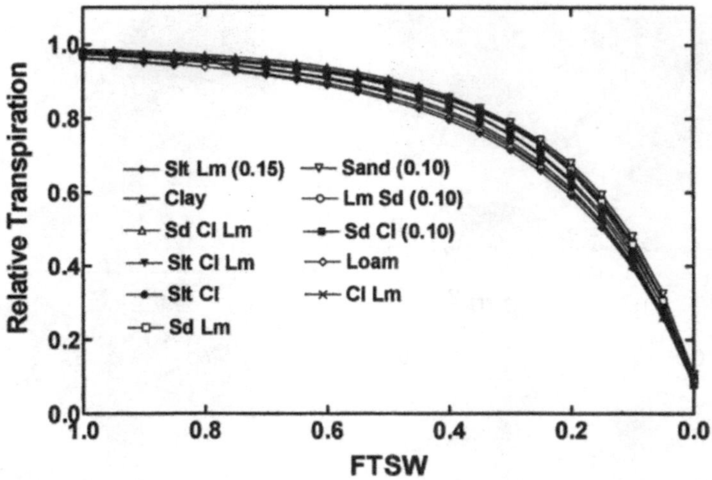

Fig. 2.2 Derived transpiration rate at various fraction transpirable soil water (FTSW) contents based on changes in soil conductivity with soil drying (Sinclair 2005)

for several soils. In these calculations there is a nonlinear decrease in relative transpiration, but the major change is initiated at about 1/3 FTSW.

The basis for initiation of early stomatal closure with soil drying is not resolved. An obvious hypothesis is that those genotypes with early closure may have low plant hydraulic conductivity resulting in a combined low soil conductivity and low plant conductivity causing the threshold for the initiation of decrease in stomatal conductance to occur at higher soil water content. This hypothesis was not supported, however, by measurements of plant conductivity of well-watered sorghum plants. Choudhary and Sinclair (2014) found those genotypes that had thresholds for initiation of stomatal closure at higher soil water content actually had higher plant hydraulic conductivity. However, they did not test for possible differences in decreases in plant conductivity among genotypes as the soil dried, which is a possible explanation for genetic differences in the threshold for decrease in stomatal closure.

2.2 Consequences of Early Stomatal Closure

Partial stomatal closure with soil drying will, of course, result in restricted plant gas exchange. The immediate negative consequence of partial closure is a decrease in CO_2 assimilation with a long-term possibility of yield decrease. The positive consequence of partial stomatal closure is that transpiration rate is decreased so that there is conservation of soil water. The initiation of soil water conservation will better position plants to sustain physiological activity over a longer time period as drought develops than those plants that delay stomatal closure until a lower FTSW is

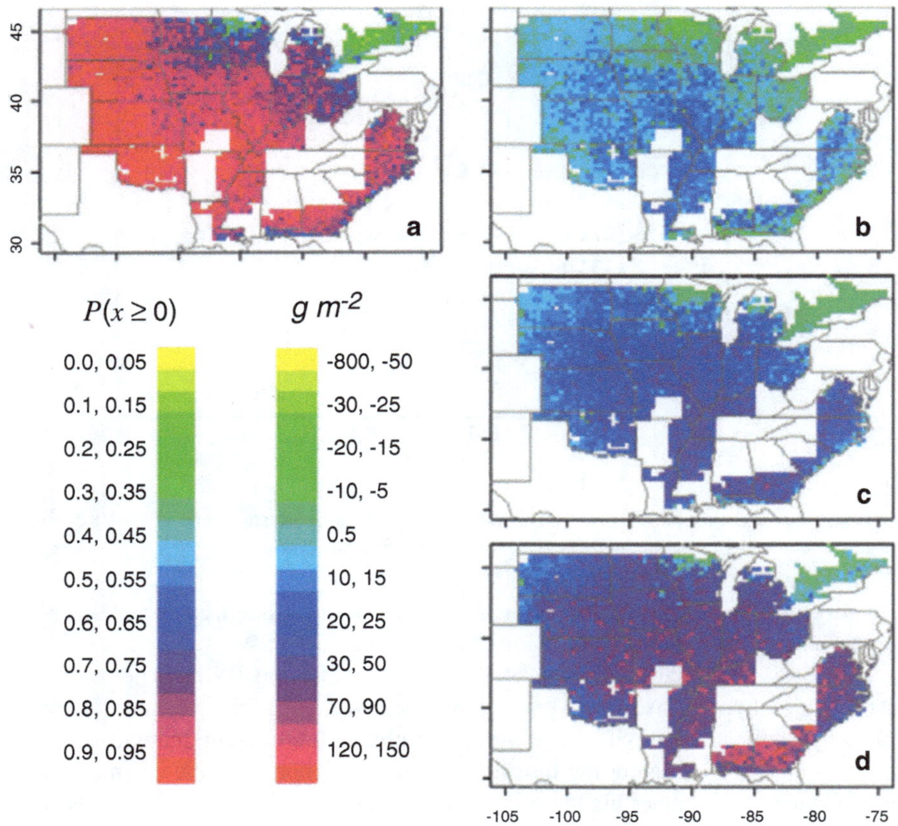

Fig. 2.3 Simulated yield increases for early partial stomatal closure of soybean with soil drying as compared to delayed partial stomatal closure. Panel (**a**) and the left-hand *color bar* are the probability of yield increase with early partial stomatal closure. Panels (**b**–**d**) and the right-hand *color bar* are the yield increases in the 75% (wet season), median, and 25% (dry season) percentile rankings (Sinclair et al. 2010)

reached. If the period of water deficit is sufficiently long, the plant genotype with partial stomatal closure as the soil dries will be better positioned to have greater integrated CO_2 assimilation and to produce greater yield. Of course, the possible occurrence of yield increase as a result of early stomatal closure depends on the FTSW at which the early closure is initiated and the duration of the drought.

To examine the possible influence on the yield of introducing early partial stomatal closure in soybean cultivars across the USA, Sinclair et al. (2010) did a simulation study using the simple simulation model (SSM). SSM is a mechanistic model that simulates in daily time steps plant development as a function of temperature, growth as a function of intercepted solar radiation, and transpiration as a function of growth. Two sets of simulations were done comparing the yields with the initiation of partial stomatal closure for FTSW of 0.21 and FTSW of 0.33. Results were obtained across 50 growing seasons for each of 2655 grid locations across the

USA. At each location, the yield difference between simulations with higher FTSW threshold minus the results with the lower FTSW threshold was calculated for each growing season and location. The results of the simulations (Fig. 2.3a) showed a yield increase due to an increased FTSW threshold in 80% or more of the growing seasons for most of the USA, including the major regions of soybean production.

The absolute yield increases obtained from simulations with the two FTSW values for the decrease in transpiration rate were also calculated. At each location, the yield difference across the 50 years was ranked and plotted for the 75% ranking (wetter season), median ranking, and 25% ranking (drier season) and is presented in Figs. 2.3b–d, respectively. For the drier ranking seasons, yield increases were generally 50–70 g m^{-2}. Even in the median yield seasons, the yield increases due to the early stomatal closure were 20–40 g m^{-2}. In the wetter season when yield loss is less likely, there were still yield increases generally in the range of 0–20 g m^{-2}. Only in the coastal and northern regions of the USA was a yield loss predicted in the wetter seasons. On balance, the early partial closure of stomata was simulated to be a candidate for increasing crop yields, especially during the drier years.

Experimental efforts to identify genotypes with early partial stomatal closure with soil drying are presented in the following chapters for most crop species. The focus of these studies was to identify genotypes with partial stomatal closure at sufficiently high soil water contents, i.e., FTSW, to be a resource in developing drought-tolerant genotypes.

References

Bennett JM, Sinclair TR, Muchow RS, Costello SR (1987) Dependence of stomatal conductance on leaf water potential, turgor potential, and relative water content in field-grown soybean and maize. Crop Sci 27:984–990

Choudhary S, Sinclair TR (2014) Hydraulic conductance differences among sorghum genotypes to explain variation in restricted transpiration rates. Funct Plant Biol 41:270–275

Kramer PJ (1944) Soil moisture in relation to plant growth. Bot Rev 10:525–559

Martin EV (1940) Effect of soil moisture on growth and transpiration in *Helianthus annuus*. Plant Physiol 14:449–466

Ritchie JT (1981) Water dynamics in the soil-plant-atmosphere system. Plant Soil 28:81–96

Sadok W, Sinclair TR (2011) Crops yield increase under water-limited conditions: review of recent physiological advances for soybean genetic improvement. Adv Agron 113:325–349

Sadras VO, Milroy SP (1996) Soil-water thresholds for the responses of leaf expansion and gas exchange: a review. Field Crops Res 47:253–266

Sinclair TR (2005) Theoretical analysis of soil and plant traits influencing daily plant water flux on drying soil. Agron J 97:1148–1152

Sinclair TR, Messina CD, Beatty A, Samples M (2010) Assessment across the United States of the benefits of altered soybean drought traits. Agron J 102:475–482

Chapter 3
Limited-Transpiration Rate Under Elevated Atmospheric Vapor Pressure Deficit

Thomas R. Sinclair

3.1 Introduction

The driving force of transpiration rate (TR) is the gradient in vapor pressure between the dry atmosphere and the wet interior of leaves, commonly referred to as the vapor pressure deficit (VPD). The other key variable is the resistance to gas diffusion attributed to the aperture of the stomata pores (r_s) and to the resistance in boundary layer of the air surrounding leaves (r_a).

$$\mathrm{TR} = \frac{\mathrm{VPD}}{(r_s + r_a)}$$

Since r_s is also a resistance to CO_2 flux into leaves to support photosynthesis, a low r_s is likely to allow greater photosynthesis rate. However, this is at the cost of greater transpiration rate. In fact, in most crop species, a sustained low r_s seems to have been favored in the selection of commercial varieties.

One approach that has not been directly explored in developing genotypes for water-deficit conditions is to identify genotypes with increasing r_s at elevated VPD. Increasing resistance at high VPD results in lower TR than would otherwise occur so that at midday, when VPD is high, there may be conservation of soil water. That is, TR does not continually increase with elevated VPD, which is especially prevalent during midday, and the total water loss by the crop is lower than would otherwise occur. The conserved soil water, especially early in the growing season, will be available to sustain crop physiological activity later in the growing season when water deficits are likely to develop in many cropping environments. Therefore, the

T.R. Sinclair (✉)
Crop and Soil Sciences Department, North Carolina State University, Raleigh, NC, USA
e-mail: trsincla@ncsu.edu

T.R. Sinclair (ed.), *Water-Conservation Traits to Increase Crop Yields in Water-deficit Environments*, SpringerBriefs in Environmental Science,
DOI 10.1007/978-3-319-56321-3_3

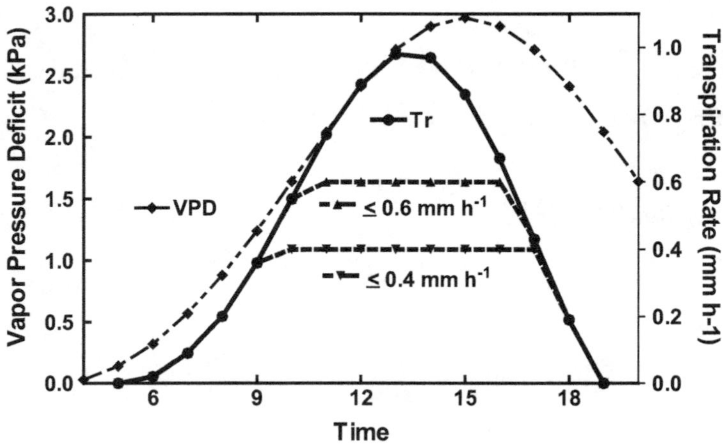

Fig. 3.1 Illustration of change through the day in vapor pressure deficit (*VPD*), transpiration rate (*TR*), and two cases of limited transpiration (Sinclair et al. 2005)

conserved water has the possibility of contributing to prolonged seed fill and yield increase. An incidental consequence of this limited-transpiration response is an increase in water-use efficiency, i.e., increased accumulated mass per unit TR and increase seed yield per unit TR.

The limited-transpiration trait is illustrated in Fig. 3.1. The dashed line with diamonds illustrates the daily change in VPD through the day with a value near zero at sunrise and a peak value in the afternoon at about 15:00. The solid line in Fig. 3.1 illustrates the transpiration rate that might be expected through the daily cycle without the limited-transpiration trait with a peak value occurring during the early afternoon. Two cases of limited-transpiration rate are shown in the figure: 0.6 and 0.4 mm h^{-1}. The integrated difference between the limited-transpiration cases and the normal transpiration rate represents the conserved soil water that results from expression of the limited-transpiration trait.

3.2 Benefit of Limited-Transpiration Trait

Since partial closure of the stomata under elevated VPD as a result of the limited-transpiration trait will also result in a decrease in photosynthesis rate, it is not clear whether the trait will actually result in yield increase. That is, will the immediate loss in photosynthetic activity under high VPD be more than compensated by sustained physiological activity later in the season as a result of the conserved soil water? Clearly, if no late-season water deficit develops, there would be no benefit from the conserved soil water, and a decrease in yield would be expected. The resolution of the possible benefit of the limited-transpiration trait requires an analysis of the temporal dynamics of the daily VPD environment and the rainfall pattern

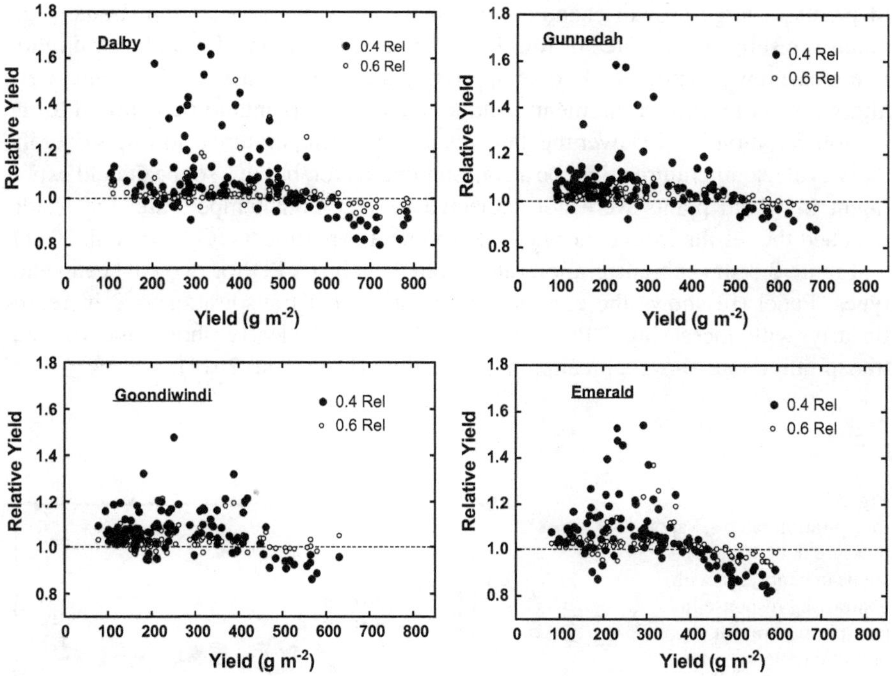

Fig. 3.2 Ratio of sorghum yield between that obtained with the limited-transpiration trait and that without the trait. Simulation results are for four locations and the two levels of limited transpiration (Sinclair et al. 2005)

through the crop growing season. Sinclair et al. (2005) did such an analysis for sorghum (*Sorghum bicolor* L.) growth and yield at four locations in Australia over more than 100 years. As shown in Fig. 3.2, there was generally a simulated yield increase (relative yield > 1.0) in the lower yielding seasons, i.e., drier seasons. In the wetter seasons (higher yielding seasons), a slight yield decrease was obtained as could be expected. The transition from yield increase to yield decrease occurred at about the yield level of 450–500 g m^{-2}. Importantly for these four locations, the yield advantage was found to occur in about 75% of the growing seasons. Since the yield increase occurred in low-yield years when growers are more economically vulnerable, Sinclair et al. (2005) suggested that this trait would be a desirable plant improvement.

3.3 Genetic Variation

Expression of the limited-transpiration trait in individual genotypes can be measured by exposing plants to a range of VPD and measuring their transpiration rate. The test can be done by enclosing plants in individual chambers or placing a group of plants in large growth chambers in which humidity can be fixed. Usually the plants are subjected to three or four levels of humidity in one day and transpiration rate measured gravimetrically over approximately 1 h for each VPD treatment. An alternative is to do measurements outdoors and allow plants to be exposed to the natural variation in VPD over the daily cycle. Since temperature also varies over the daily cycle, temperature might be a confounding factor, but in a soybean field experiment in which plants were not subjected to very high temperature, the results matched the results from controlled environment experiments (Gilbert et al. 2011).

Figure 3.3 shows transpiration rates plotted against VPD for two soybean genotypes. Panel (b) shows the common observation that transpiration rate increases linearly with increasing VPD. Genotype N05-7432 clearly shows the limited-transpiration rate above a breakpoint at 1.47 kPa (Devi et al. 2014).

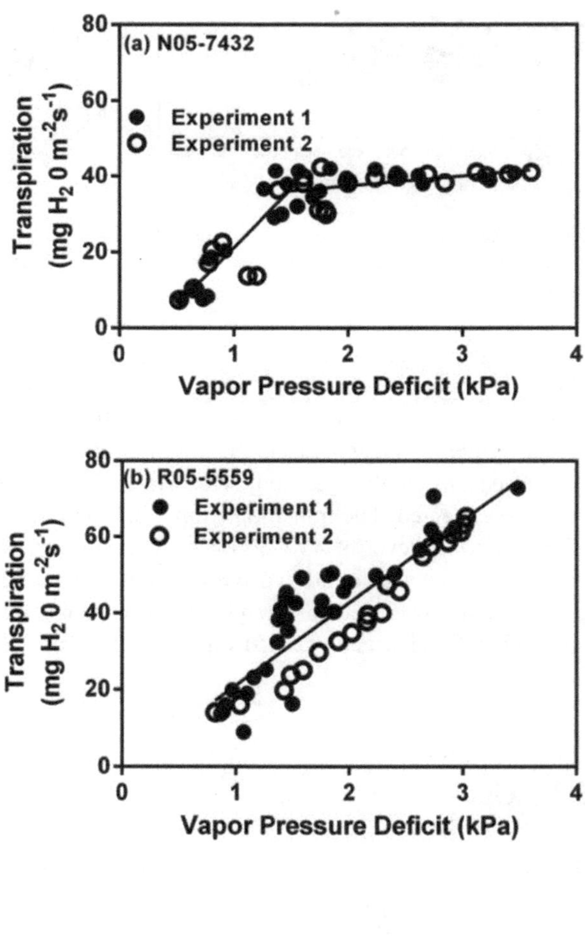

Fig. 3.3 Plot of transpiration rate vs. vapor pressure deficit for two soybean genotypes with contrasting response in transpiration rate at vapor pressure deficit greater than 1.47 kPa (Devi et al. 2014)

3.4 Mechanistic Basis for Limited Transpiration

The mechanistic basis for expression of the limited-transpiration trait is not fully resolved, but a limited hydraulic conductivity (inverse of resistance) in the leaves or elsewhere has been suggested (Sinclair et al. 2017). A low hydraulic conductivity means that water cannot move as rapidly to the leaves, and under high VPD this makes leaves, or at least individual cells, vulnerable to dehydration. Therefore, partial stomatal closure is required to match transpiration rate with water flux to the leaves to avoid leaf desiccation and senescence (Bunce 2006). In soybean, Sinclair et al. (2008) reported that genotype PI 416937, which expresses the limited-transpiration trait, had lower hydraulic conductivity in the leaves as compared to the other tested genotypes. The site of the lower conductivity was identified as being in the water pathway in the leaves between the xylem and stomata.

It has been proposed that the main location of resistance to water flow are transmembrane proteins, or aquaporins, in the water flow pathway (Heinen et al. 2009). Studies with aquaporin inhibitors have shown segregation in expression of limited transpiration following treatment of leaves (Sadok and Sinclair 2010a, b). Silver ions, in particular, seem effective in resulting in discrimination among genotypes. The genotypes that express the limited-transpiration trait are insensitive to the silver treatment, which may indicate that these genotypes do not have the population of silver-sensitive aquaporins that facilitate higher conductivity in the "normal" genotypes. The discrimination among genotypes in silver sensitivity has also been shown in peanut (Devi et al. 2012; Shekoofa et al. 2013), sorghum (Choudhary et al. 2013), and maize (Choudhary et al. 2015).

As discussed in the case studies in this book, the limited-transpiration trait has been a focus of considerable research in many crop species. This research has ranged from efforts to identify genotypes within a species that express the trait to understanding the basic mechanisms in the plant that result in limited transpiration. Now, commercial hybrids of maize are available (AQUAmax, DuPont Pioneer), and it appears commercial varieties of soybean may also be available (Devi et al. 2014). A major part of this book is to provide reviews of the various research that has been undertaken on the limited-transpiration trait.

References

Bunce JA (2006) How do leaf hydraulics limit stomatal conductance at high water vapour pressure deficits? Plant Cell Environ 29:1644–1650

Choudhary S, Sinclair TR, Prasad PV (2013) Hydraulic conductance of intact plants of two contrasting sorghum lines, SC15 and SC1205. Funct Plant Biol 40:730–738

Choudhary S, Sinclair TR, Messina CD, Cai W, Warner D, Cooper M (2015) Inhibitor screen for limited-transpiration trait among maize hybrids. Environ Exp Bot 109:161–167

Devi MJ, Sadok W, Sinclair TR (2012) Transpiration response of de-rooted peanut plants to aquaporin inhibitors. Environ Exp Bot 78:167–172

Devi MJ, Sinclair TR, Chen P, Carter TE (2014) Evaluation of elite southern maturity soybean breeding lines for drought-tolerant traits. Agron J 106:1947–1954

Gilbert ME, Holbrook NM, Zwieniecki MA, Sadok W, Sinclair TR (2011) Field confirmation of genetic variation in soybean transpiration response to vapor pressure deficit and photosynthetic compensation. Field Crop Res 124:85–92

Heinen RB, Ye Q, Chaumont F (2009) Role of aquaporins in leaf physiology. J Exp Bot 60:2971–2985

Sadok W, Sinclair TR (2010a) Transpiration response of 'slow-wilting' and commercial soybean (*Glycine max* (L.) Merr.) genotypes to three aquaporin inhibitors. J Exp Bot 61:821–829

Sadok W, Sinclair TR (2010b) Genetic variability of transpiration response of soybean (*Glycine max* (L.) Merr.) shoots to leaf hydraulic conductance inhibitor $AgNO_3$. Crop Sci 50:1423–1430

Shekoofa A, Devi JM, Sinclair TR, Holbrook CC, Isleib TG (2013) Divergence in drought-resistance traits among parents of recombinant peanut inbred lines. Crop Sci 53:2569–2576

Sinclair TR, Devi J, Shekoofa A, Choudhary S, Sadok W, Vadez V, Riar M, Rufty TW (2017) Limited-transpiration response to high vapor pressure deficit in crop species. Plant Sci (in press)

Sinclair TR, Hammer GL, van Oosterom EJ (2005) Potential yield and water-use efficiency benefits in sorghum from limited maximum transpiration rate. Funct Plant Biol 32:945–952

Sinclair TR, Zwieniecki MA, Holbrook NM (2008) Low leaf hydraulic conductance associated with drought tolerance in soybean. Physiol Plant 132:446–451

Chapter 4
Soybean

Thomas R. Sinclair

The seeds of soybean (*Glycine max* (Merr.) L) are high in both protein and oil concentration. As such, the seeds are a valuable source of food and feed protein, cooking oil, and many industrial uses. A major advantage of soybean is that it has high symbiotic nitrogen fixation capacity, so nitrogen fertilization is not required. Given the continually increasing demand and no need to apply nitrogen fertilizer, the land area committed to soybean production has steady increased over the past several decades. The major producing countries are the USA, Brazil, Argentina, India, and China. Soybean is not often irrigated (<5% of the land area in the major producing countries), so the crop is vulnerable to variations in rainfall. Even short periods of soil water deficit can adversely affect soybean yields, especially because of the high sensitivity of its nitrogen fixation to even minor decreases in soil moisture. Hence, mechanisms that allow maintenance of soil water content at higher levels are likely to be beneficial in increasing soybean yields.

4.1 Water Conservation with Soil Drying

An early investigation of the relationship for soybean between transpiration rate and volumetric soil water content was presented by Sinclair and Ludlow (1986). They presented the results of experiments with a single cultivar done in pots containing sandy loam soil using a logistic equation. Based on this equation, the fraction of transpirable soil water (FTSW) remaining in the drying soil when transpiration rate was 90% of well-watered plants was 0.21. Serraj and Sinclair (1997) compared

T.R. Sinclair (✉)
Crop and Soil Sciences Department, North Carolina State University, Raleigh, NC, USA
e-mail: trsincla@ncsu.edu

T.R. Sinclair (ed.), *Water-Conservation Traits to Increase Crop Yields in Water-deficit Environments*, SpringerBriefs in Environmental Science,
DOI 10.1007/978-3-319-56321-3_4

eight genetically divergent soybean cultivars, and all showed initially a plateau in transpiration rate with drying soil followed by a decreasing rate. All their results were represented by a single exponential equation, and the 90% level in the decline of transpiration rate occurred at FTSW of 0.34. Sadok et al. (2012) compared six diverse genotypes, and the threshold decline in transpiration as represented by a two-segment linear model was in the fairly narrow FTSW range of 0.25–0.32. Devi et al. (2014) showed no difference in the threshold for transpiration decline among ten soybean genotypes, although the threshold values were high in their experiment as compared to other experiments.

Hufstetler et al. (2007) investigated a total of 24 soybean genotypes from a range of maturity groups. One study was done on a sandy soil (80% sand), and they found the FTSW ranged from 0.22 to 0.29 for the 12 tested genotypes. In the second study on artificial potting medium, the FTSW ranged from 0.19 to 0.27 for a second set of genotypes. The thresholds in this experiment were likely lower than observed in other experiments due to the coarse nature of the potting medium used. Sinclair et al. (1998) showed that FTSW for sandy soils resulted in unusually low thresholds for the decline in transpiration rate. Nevertheless, the results of Hufstetler et al. (2007) do not offer a great deal of support for exploitable genetic variation in the threshold for decline in transpiration rate with soil drying.

Ray and Sinclair (1998) examined the possible influence of pot size on experimental results in the transpiration response to soil drying for soybean. Pot size did not impact the response, and the decline in transpiration rate of the stress plants to 90% of the well-watered plants based on their exponential result was at FTSW equal to 0.33. The threshold for the decrease in transpiration rate was also found to be insensitive to carbon dioxide treatment by Serraj and Sinclair (1999) with a common 90% threshold response pattern for two carbon dioxide treatments.

Thus far, experimental evidence has not resulted in the identification of a soybean genotype with an unusually high FTSW threshold for stomatal closure in response to soil drying. Nevertheless, simulation studies as presented in Chap. 2 indicate a very favorable impact on yield in the USA if early stomatal closure was introduced into soybean cultivars. One point of optimism that a soybean genotype expressing a high threshold for initiation of transpiration decrease might yet be discovered or developed comes from the study of ancestral soybean (*Glycine soja*) by Seversike et al. (2014). When *Glycine soja* plants were grown at 30 °C, the threshold for the decline in transpiration was at FTSW 0.28. However, when the plants were grown at 25 °C, the FTSW threshold for the decline in transpiration rate increased to 0.67. The basis for this change in threshold with temperature is unknown, but one possibility might be a change in the hydraulic conductivity of the plants. If the shift to a low conductivity occurred at 25 °C in *Glycine soja* plants, then it seems possible there may be a *Glycine max* genotype with a low hydraulic conductivity even at higher temperatures. The discovery of such a *Glycine max* genotype, or its development using *Glycine soja* germplasm, could introduce an important trait for drought tolerance in soybean.

4.2 Water Conservation by Sensitivity to Vapor Pressure Deficit

4.2.1 Slow-Wilting Phenotype

An early observation of soybean drought response in the field was with genotype PI 416937, which originated from Japan but has unknown parentage. This genotype had visually delayed wilting under drought conditions as compared to nearly all other tested lines (Sloane et al. 1990). Based on these field observations, PI 416937 was labeled as "slow-wilting." Although initial observations resulted in the suggestion that the advantage of this genotype was prolific fibrous rooting in the upper soil zone (Hudak and Patterson 1995), such a trait was not shown to result in greater access to soil water, and this hypothesis has been rejected (King et al. 2009).

4.2.2 Observations of Water-Limitation Trait

An alternative to rooting traits to explain the slow-wilting phenotype was soil water conservation under conditions of high atmospheric vapor-pressure deficit (VPD). Water conservation resulting from such a limited-transpiration trait under high VPD would allow more water to be left in the soil for use later in the growing season. The conserved soil water would allow sustained physiological activity during the critical period of reproductive development and seed growth. In fact, the possibility of partial stomatal closure under high VPD in soybean has long been recognized. Bunce (1982) found for cultivar Kent grown in the field that leaf water-vapor conductance was less on days when VPD was high as compared to days when VPD was low. In tests both in the field and controlled environment, Bunce (1984) found that the ratio of water-vapor conductance between treatments of leaf to air VPD of 3.0 vs. 1.5 kPa was less than 1.0 for each of the three tested soybean cultivars. Genetic differences among the three cultivars were identified in this study in the degree of decrease in stomatal conductance under the high VPD treatment.

Examination of transpiration rate of soybean plants over a wide range of VPD was not reported until the study of Fletcher et al. (2007). They compared the response in transpiration rate to VPD in a controlled environment for four genotypes. One of the genotypes, slow-wilting PI 416937, exhibited a breakpoint in its transpiration response to VPD such that above VPD of 2.13 kPa; there was no further increase in transpiration rate. This contrasted with the other three genotypes, which exhibited a continuing linear increase in transpiration rate with increasing VPD.

Gilbert et al. (2011) found the limited-transpiration response above a threshold VPD, for PI 416937 was also expressed in the field. By measuring stomatal conductance over naturally varying VPD through the day, Gilbert et al. (2011) observed a breakpoint in the response for PI 416937 at about 2 kPa. Stomatal conductance continued to decline with increasing VPD above the breakpoint resulting in stable transpiration rate at VPD greater than the breakpoint. Included in their study were six other genotypes selected for expression of the limited-transpiration trait based on controlled-chambers studies, and these were all confirmed to have expression of the trait in the field.

Sinclair et al. (2008) not only confirmed the transpiration response of PI 416937 to VPD but found that this genotype had low hydraulic conductance in its leaves as compared to two other genotypes. To explain this difference, they proposed the hypothesis that low leaf hydraulic conductance restricts water flux to stomata, and as a result, stomata will partially close due to an inability to rapidly replenish epidermal water levels as rapidly as other genotypes. Evidence for the low leaf hydraulic conductance of PI 416937 was obtained in three ways. (1) The rehydration of partly dried leaves was measured and characterized by the time constant for rehydration. The time constant for rehydration of PI 416937 was significantly longer than the other two tested genotypes indicating lower hydraulic conductance. (2) The rhizosphere of plants was pressurized, and transpiration rate was measured. While the slope of the transpiration response above the breakpoint was increased in two genotypes as a result of the pressurization, the slope for PI 416937 was unchanged. These results indicated that there was a lower hydraulic conductance in PI 416937 that was not greatly compensated by the pressurization treatment. (3) Gas exchange of leaves was measured after cutting leaf petioles under water. This treatment was expected to allow an immediate surge of water into epidermal cells that would result in an increase in the epidermal cell turgor pressure as compared to the guard cells, so there would be a decrease in leaf gas exchange. This response was indeed observed, and the decrease in transpiration rate in PI 416937 was slower and to a lower extent as compared to two other genotypes. Therefore, all three approaches were consistent in indicating a lower hydraulic conductance in PI 416937 as compared to the other two tested genotypes.

4.3 Aquaporin Involvement

No morphological features of the leaves of PI 416937 have been identified that explain its lower hydraulic conductance. Considering that a considerable amount of water is transported from the xylem to guard cells in a transmembrane pathway (Heinen et al. 2009), a physiological hypothesis to explain a lower hydraulic conductance is that aquaporins (AQP) embedded in the membranes of PI 416937 might result in a lower leaf hydraulic conductance. As a possible means of showing the involvement of AQP, experiments were done to observe the transpiration rate of de-rooted shoots after being fed various mercurial metals, which are commonly observed to inhibit AQP expression. Indeed, feeding of mercury resulted in up to a 75% decrease in transpiration rate (Sadok and Sinclair 2010a) in all genotypes including PI 416937.

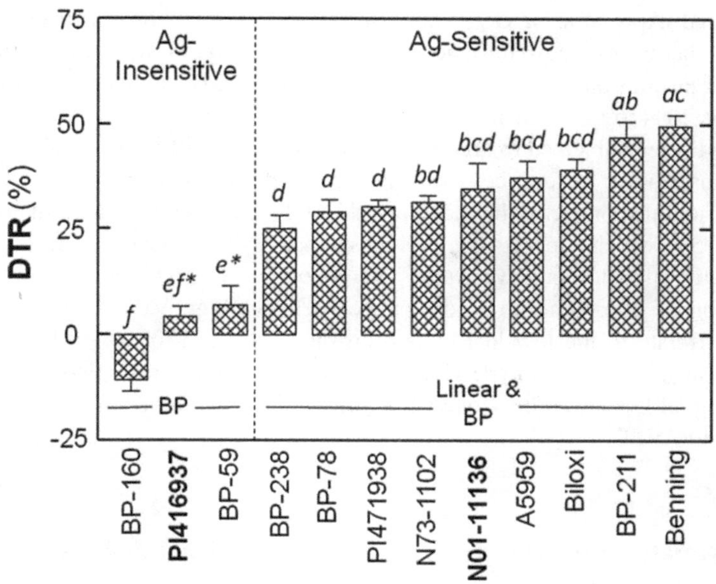

Fig. 4.1 Percent decrease in transpiration rate (*DTR*) after exposure of de-rooted shoots to silver ion. Genotypes noted by *different letters* were significantly different ($p < 0.05$) (Sadok and Sinclair 2010b)

However, when de-rooted shoots were fed silver ions, there was differential response among genotypes in their transpiration rate (Sadok and Sinclair 2010a). The transpiration rate of PI 416937 was nearly insensitive to the feeding of silver ions, while the other tested genotypes had about a 40% decrease in transpiration rate. Subsequently, Sadok and Sinclair (2010b) found a large range among genotypes in response to silver treatment with PI 416937 and two progeny lines being essentially insensitive to the silver treatment, while the remaining genotypes showed sensitivity (Fig. 4.1). A hypothesis to explain this observation is that PI 416937 has a population of AQP that are not sensitive to silver, while most other genotypes have a population of AQP that are sensitive to silver and hence, show a decreased transpiration rate with the silver treatment. The advantage of those genotypes with silver-sensitive AQP was proposed to be a capacity to more readily transport water, i.e., a higher hydraulic conductance, compared to PI 416937 in which silver-sensitive AQPs are absent or have a much lower abundance.

Sadok and Sinclair (2012) also found that treatment of de-rooted soybean shoots with zinc ions also resulted in segregation among soybean genotypes. In this case, transpiration rate of PI 416937 following the feeding of de-rooted shoots with zinc resulted in large decreases in transpiration rate as compared to other genotypes. This result offers further evidence of a differing population of AQP in PI 416937, which appear to be especially sensitive to exposure to zinc.

Recently, AQP transcript abundance studies in soybean following treatment with AQP inhibitors silver and zinc confirmed the likely interaction of AQPs with metals and altered expression of AQP transcripts (Devi et al. 2016). A difference was observed between gene expression of soybean genotypes with and without the limited-transpiration trait when subjected to high VPD compared to low VPD. Genotype PI 416937 with limited transpiration showed 944 genes differentially expressed when subjected to high VPD compared to low VPD. Most of these genes were transcription factors, cell wall development and metabolism, and transport genes. The transport genes included mostly AQPs that were down regulated (Devi et al. 2015a), which supports the hypothesis that AQPs are involved in the limited-transpiration trait to conserve water. The genotype PI 416937 also showed lower expression of expansin and extension genes under high VPD conditions which might correlate to the low leaf expansion rate compared to low VPD conditions (Devi et al. 2015b). Genotype Hutcheson with increasing transpiration rate with increasing VPD showed only one gene differentially expressed at high VPD compared to low VPD.

4.4 Genetic Variation for Limited-Transpiration Trait

There have been several studies to investigate genetic variation among soybean genotypes for expression of the limited-transpiration trait. Sadok and Sinclair (2009a) examined seven breeding lines in which four had PI 416937 in their pedigrees. Interestingly, none of the four lines derived from PI416937 exhibited the limited-transpiration trait, but two of the other genotypes expressed the trait. To better document the potential for transferring the limited-transpiration trait to progeny lines, Sadok and Sinclair (2009b) measured expression of the limited-transpiration trait in a controlled environment for 22 progeny recombinant inbred lines (RIL) derived from the mating of PI 416937 and Benning. Thirteen of the RILs did not express the limited-transpiration trait further indicating the trait might be readily lost in breeding efforts. However, eight RILs expressed the trait and had thresholds for limited transpiration well below either of the parents (1.1–1.5 kPa vs. ~2.1 kPa). One RIL had a breakpoint similar to the parental lines.

The results of Sadok and Sinclair (2009a, b) indicated that the ability to transfer expression of the limited-transpiration trait to progeny lines is complex. However, the laborious methodology to directly phenotype for transpiration response to VPD has not yet allowed a direct assessment of classical inheritance. To gain preliminary insight about inheritance, Carpentieri-Pipolo et al. (2012) attempted to identify QTL markers for the limited-transpiration trait based on a silver-feeding screen, which was used as a surrogate for the limited-transpiration trait. They phenotyped 147 recombinant inbred lines (RIL) derived from the mating of PI 416937 with Benning, which also expresses the limited-transpiration trait but is not silver sensitive (Sadok and Sinclair 2009b). Four SSR markers, two associated with each parent, were identified that were correlated with insensitivity to silver treatment, each accounting for 16% or more of the variation. These markers may offer a useful tool in tracking the limited-transpiration trait in soybean populations.

4.5 Germplasm Release

Several soybean germplasm releases with PI 416937 in their pedigree have been developed that produced high yields in classical yield trials (Carter et al. 2003, 2007, 2008). The most recent release has been a cultivar release specifically for drought resistance (Carter et al. 2016). Seed of these releases have been widely distributed, and commercial cultivars have now been found that express the limited-transpiration trait (T.R. Sinclair, personal communication).

Breeding efforts have continued to use PI 416937 as a parent for developing new genotypes with drought tolerance. Devi et al. (2014) tested for expression of the water-limited train in ten genotypes expressing high yields in yield trials and were top candidates for cultivar or germplasm release from the breeding programs at the University of Arkansas and USDA-ARS, Raleigh, NC. All but one of these ten lines was found to express the limited-transpiration trait with breakpoints ranging from 1.4 to 1.9 kPa. Interestingly, only five of the nine lines had PI 416937 in their pedigree. The potential source(s) of the limited-transpiration trait of the other four lines is unknown. Nevertheless, these results indicated that those lines expressing the limited-transpiration traits were at a competitive advantage in achieving high yields.

4.6 Extent of Yield Advantage

While selection in a breeding program over the test environments showed a yield advantage for the limited-transpiration trait, the geographical extent of the advantage and probability of yield increase could not be resolved from such limited experimental data. The SSM model for soybean was used to answer these questions for the USA and Africa. To simulate the limited-transpiration trait, the daily weather input data were expanded to hourly values by using standard diurnal functions to estimate the hourly weather data from daily values. In the US simulations for the limited-transpiration trait, whenever hourly VPD was greater than 2 kPa, transpiration rate was set equal to the rate at the threshold, and hourly growth was adjusted accordingly (Sinclair et al. 2010).

The results of 50-y simulations for 2655 locations in the USA are shown in Fig. 4.2. Panel [4.2a] shows the results for the probability of yield increase at each location across the simulated years. In fact, the probability of yield gain was 85% or greater in nearly all locations in the USA, including regions in the Central Midwest that are commonly assumed not to be water limited. These simulation results indicated that in a large number of years at nearly all locations, the crop suffered at least brief, intermittent drought periods that can be ameliorated to some extent with the limited-transpiration trait. Panels [4.2b–d] show the absolute yield gain when the yields at each locations are ranked across growing season and the results for the limited-transpiration trait at the 25%, median, and 75% percentile ranking, where the 25% percentile ranking represents the drier, lower yielding

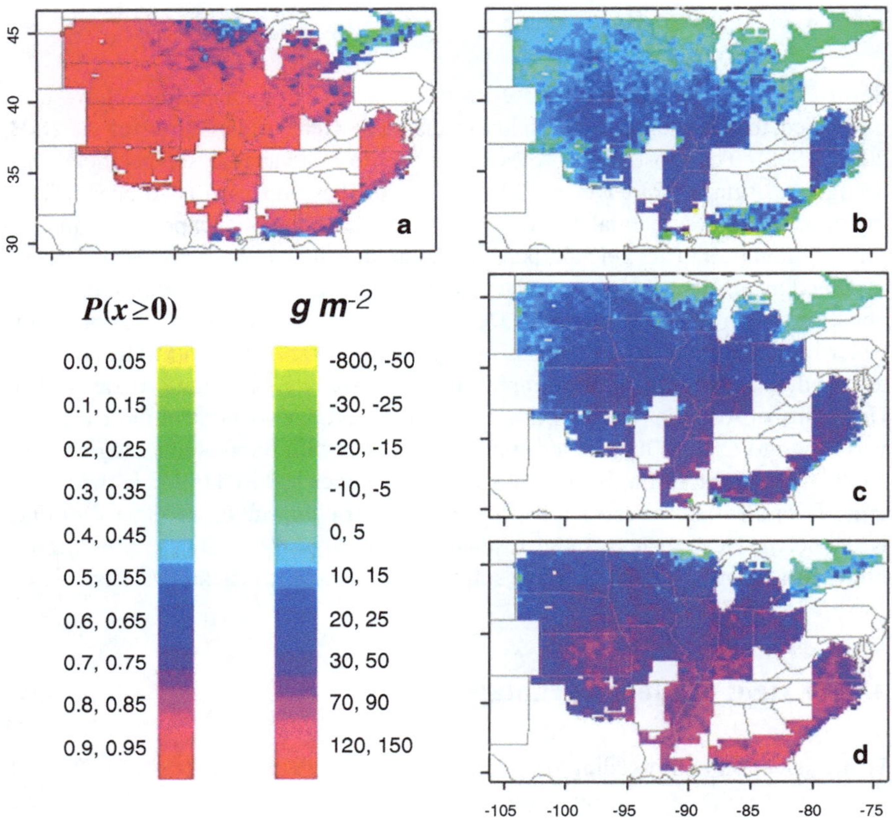

Fig. 4.2 Results of simulations of soybean across the USA with inclusion of the limited-transpiration trait across the USA (Sinclair et al. 2010). (**a**) Probability of yield increase with limited-transpiration trait as compared to soybean without the trait. (**b–d**) Yield difference between the simulation with limited-transpiration trait and standard simulation was calculated for each growing season and grid location. The yield differences were ranked across growing seasons at each grid location from highest to lowest and the yield difference at the (**b**) 75%, (**c**) median, and (**d**) 25% percentile are presented. Simulations with SSM were done using 50-y of weather data. The limited-transpiration trait was imposed in the simulations using a threshold of 1.8 kPa (Sinclair et al. 2014)

growing season. In the 25% percentile season, yields tended to be increased substantially in the range of 50–100 g m^{-2} in most locations. Even in the wet seasons, i.e., 75% percentile, the yield increase was commonly in the range of 20–30 g m^{-2}. Hence, the simulations showed the limited-transpiration trait has a broad geographical yield advantage, and the yield advantage was simulated for a preponderance of growing season.

A simulation study of soybean yield similar to the one in the USA was done for Africa (Sinclair et al., 2014). Weather data was generated for 50 years in both west and east Africa. Due to the generally drier conditions, background yield were commonly low with yields in individual seasons rarely greater than 300 g m^{-2}.

The limited-transpiration trait was imposed in the simulations using a threshold of 1.8 kPa. The average yield change due to the limited-transpiration trait was greater than 15 g m^{-2} for the northern regions of west Africa and many locations in east Africa. The probability of yield increase was 85% or greater in these regions. Therefore, the simulations identified specific geographical regions in Africa where the limited-transpiration trait would be especially advantageous for increasing soybean production.

References

Bunce JA (1982) Photosynthesis at ambient and elevated humidity over a growing season in soybean. Photosyn Res 3:307–311

Bunce JA (1984) Identifying soybean lines differing in gas exchange sensitivity to humidity. Ann Appl Biol 105:313–318

Carpentieri-Pipolo V, Pipolo AE, Abdel-Haleem H, Boerma HR, Sinclair TR (2012) Identification of QTLs associated with limited leaf hydraulic conductance in soybean. Euphytica 186: 679–686

Carter TE Jr, Burton JW, Bowman DT, Cui Z, Zhou X, Villagarcia MR, Niewoehner AS, Fountain MO (2003) Registration of 'N7001' soybean. Crop Sci 43:1126–1127

Carter TE Jr, Burton JW, Fountain MO, Rzewnicki PE, Villagarcia MR, Bowman DT (2007) Registration of 'N7002' soybean. J Plant Reg 1:93–94

Carter TE Jr, Burton JW, Fountain MO, Rzewnicki PE, Villagarcia MR, Bowman DT (2008) Registration of 'N8001' soybean. J Plant Reg 2:22–23

Carter TE Jr, Todd SM, Gillen AM (2016) Registration of 'USDA-N8002' soybean cultivar with high yield and abiotic stress resistance traits. J Plant Reg 10:238–245

Devi JM, Sinclair TR, Chen P, Carter TE Jr (2014) Evaluation of elite southern maturity soybean breeding lines for drought-tolerant traits. Agron J 106:1947–1954

Devi MJ, Sinclair TR, Taliercio E (2015a) Comparisons of the effects of elevated vapor pressure deficit on gene expression in leaves among two fast-wilting and a slow-wilting soybean. PLoS One 10:e0139134. doi:10.1371/journal.pone.0139134

Devi MJ, Taliercio EW, Sinclair TR (2015b) Leaf expansion of soybean subjected to high and low atmospheric vapour pressure deficits. J Exp Bot 66:1845–1850

Devi MJ, Sinclair TR, Taliercio E (2016) Silver and zinc inhibitors influence transpiration rate and aquaporin transcript abundance in intact soybean plants. Environ Exp Bot 122:168–175

Fletcher AL, Sinclair TR, Allen LH Jr (2007) Transpiration responses to vapor pressured deficit in well watered 'slow wilting' and commercial soybean. Environ Exp Bot 61:145–151

Gilbert ME, Holbrook NM, Zwieniecki MA, Sadok W, Sinclair TR (2011) Field confirmation of genetic variation in soybean transpiration response to vapor pressure deficit and photosynthetic compensation. Field Crop Res 124:85–92

Heinen RB, Ye Q, Chaumont F (2009) Role of aquaporins in leaf physiology. J Exp Bot 60:2971–2985

Hudak CM, Patterson RP (1995) Vegetative growth analysis of a drought-resistant soybean plant introduction. Crop Sci 35:464–471

Hufstetler EV, Boerma HR, Carter TE Jr, Earl HJ (2007) Genotypic variation for three physiological traits affecting drought tolerance in soybean. Crop Sci 47:25–35

King CA, Purcell LC, Brye KR (2009) Differential wilting among soybean genotypes in response to water deficit. Crop Sci 49:290–298

Ray JD, Sinclair TR (1998) The effect of pot size on growth and transpiration of maize and soybean during water deficit stress. J Exp Bot 49:1381–1386

Sadok W, Sinclair TR (2009a) Genetic variability of transpiration response to vapor pressure deficit among soybean cultivars. Crop Sci 49:955–960
Sadok W, Sinclair TR (2009b) Genetic variability of transpiration response to vapor pressure deficit among soybean (*Glycine max* [L.] Merr.) genotypes selected from a recombinant inbred line population. Field Crop Res 113:156–160
Sadok W, Sinclair TR (2010a) Transpiration response of 'slow-wilting' and commercial soybean (*Glycine max* (L.) Merr.) genotypes to three aquaporin inhibitors. J Exp Bot 61:821–829
Sadok W, Sinclair TR (2010b) Genetic variability of transpiration response of soybean (*Glycine max* (L.) Merr.) shoots to leaf hydraulic conductance inhibitor $AgNO_3$. Crop Sci 50:1423–1430
Sadok W, Sinclair TR (2012) Zinc treatment results in transpiration rate decreases that vary among soybean genotypes. J Plant Nutr 35:1866–1877
Sadok W, Gilbert ME, Raza AS, Sinclair TR (2012) Basis of slow-wilting phenotype in soybean PI 471938. Crop Sci 52:1261–1269
Serraj R, Sinclair TR (1997) Variation among soybean cultivars in dinitrogen fixation response to drought. Agron J 89:963–969
Serraj R, Sinclair TR (1999) Soybean leaf growth and gas exchange response to drought under carbon dioxide enrichment. Glob Chang Biol 5:283–291
Seversike TM, Sermons SM, Sinclair TR, Carter TE Jr, Rufty TW (2014) Physiological properties of a drought-resistant wild soybean genotype: transpiration control with soil drying and expression of root morphology. Plant Soil 374:359–370
Sinclair TR, Ludlow MM (1986) Influence of soil water supply on the plant water balance of four tropical grain legumes. Aust J Plant Phyisol 13:329–341
Sinclair TR, Hammond LC, Harrison J (1998) Extractable soil water and transpiration rate of soybean on sandy soil. Agron J 90:363–368
Sinclair TR, Zwieniecki MA, Holbrook NM (2008) Low leaf hydraulic conductance associated with drought tolerance in soybean. Physiol Plant 132:446–451
Sinclair TR, Messina CD, Beatty A, Samples M (2010) Assessment across the United States of the benefits of altered soybean drought traits. Agron J 102:475–482
Sinclair TR, Marrou H, Soltani A, Vadez V, Chandolu KC (2014) Soybean production potential in Africa. Glob Food Sec 3:31–40
Sloane RJ, Patterson RP, Carter TE Jr (1990) Field drought tolerance of a soybean plant introduction. Crop Sci 30:118–123

Chapter 5
Peanut

M. Jyostna Devi and Avat Shekoofa

Peanut (*Arachis hypogaea* L.) is an important grain legume whose yield is often limited by water deficits because it is commonly grown in regions with intermittent rainfall and on sandy soils with low water-holding capacity. Annual peanut yield loss due to drought worldwide has been estimated at US$520 million (Sharma and Lavanya 2002). To increase yields, it is essential to develop genotypes that can sustain growth and yield when subjected to water deficit (Sinclair and Vadez 2012). The two plant traits, which are the topic of this book, have been identified in peanut to result in conservative use of water and allow sustained growth during drought: (1) earlier closure of stomata in the soil drying cycle resulting in decreased daily transpiration rate and (2) limited-transpiration rate (TR_{lim}) under high atmospheric vapor pressure deficit (VPD) (Devi et al. 2009, 2010).

5.1 Water Conservation with Soil Drying Cycle

Fraction of transpirable soil water (FTSW) is used as a covariate for soil moisture available to compare TR and different physiological mechanisms to soil drying. The two-linear-segment regression successfully described the TR response to soil drying in a range of germplasm including transgenic peanuts transformed with *DREB1A* transcription factor and breeding and commercial varieties (Bhatnagar-Mathur et al.

M.J. Devi (✉)
Crop and Soil Sciences Department, North Carolina State University, Raleigh, NC, USA
e-mail: mjdevi@gmail.com

A. Shekoofa
Department of Plant Sciences, University of Tennessee Knoxville, Jackson, TN 38301, USA
e-mail: ashekoof@utk.edu

T.R. Sinclair (ed.), *Water-Conservation Traits to Increase Crop Yields in Water-deficit Environments*, SpringerBriefs in Environmental Science,
DOI 10.1007/978-3-319-56321-3_5

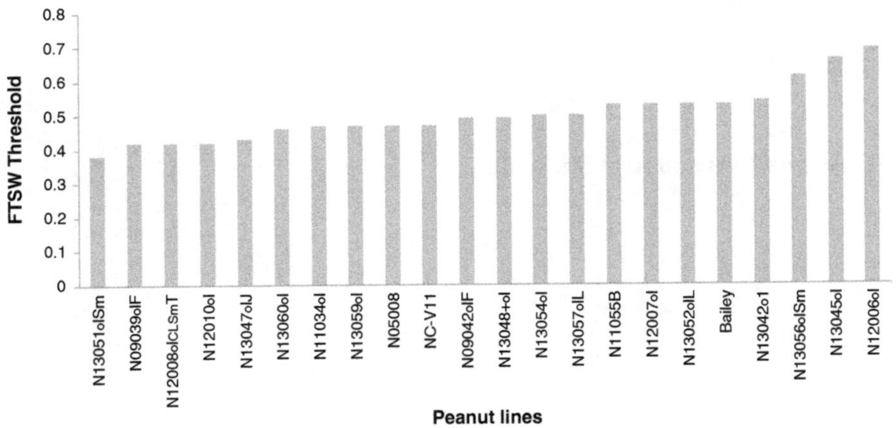

Fig. 5.1 The FTSW threshold for initiation of partial stomatal closure of Virginia-type breeding lines (Shekoofa et al. 2017b)

2007; Devi et al. 2009; Devi and Sinclair 2011; Shekoofa et al. 2013). However, the FTSW breakpoint between the plateau region of constant TR at wetter soil water and decreasing TR with soil drying has been found to be quite variable among peanut genotypes. The range of FTSW threshold to soil drying observed was between 0.36 and 0.64 for transgenic peanut (Bhatnagar-Mathur et al. 2007), 0.22–0.71 for breeding lines from India (Devi et al. 2009), and 0.36–0.46 for commercial US varieties (Devi and Sinclair 2011; Shekoofa et al. 2013).

In a recent study on 50 lines of Virginia-type peanuts for both a RIL population and breeding lines, Shekoofa et al. (2017b) found many of the lines exhibited a FTSW threshold in the range from 0.35 to 0.45 (Fig. 5.1), which is commonly observed for most plants. However, several lines expressed FTSW thresholds that were greater than this range. Three lines from the breeding population had thresholds at greater than 0.60 FTSW. The higher FTSW threshold values indicate a substantial sensitivity to soil drying, and for these genotypes a potential for soil water conservation and greater drought tolerance.

5.2 Water Conservation with Sensitivity to Vapor Pressure Deficit

5.2.1 Observations of Water-Limitation Trait

The second approach to soil water conservation is to have limited-transpiration rate (TR_{lim}) at elevated VPD such that crop water loss rate was less than normally expected. The search for the TR_{lim} in peanut has included studies in both controlled environmental conditions and field (Devi et al. 2010; Devi and Sinclair 2011; Shekoofa et al. 2013, 2015). Genotypic variation in TR response to different VPD was identified in breeding, commercial, and recombinant inbred lines (RIL) in peanut. Importantly, it was found that some genotypes had TR_{lim} with increasing VPD approximately above 2 kPa, such that there was little or no further increase in TR with increasing VPD (Fig. 5.2).

Devi et al. (2010) studied 17 breeding lines from India that were originally selected for differences in transpiration efficiency on drying soil for their transpiration rate (TR) response to VPD. It was observed that 9 out of the 17 lines expressed TR_{lim} with no or little increase in TR above about 2 kPa VPD. The remaining eight lines continued to increase their TR with increasing VPD. The threshold range of VPD for initiation of TR_{lim} was found to occur between 1.6 and 2.9 kPa for breeding lines, and the TR increase rate after the TR_{lim} breakpoint was from −6.27 to +6.34 mg $H_2O\ m^{-2}\ s^{-1}\ kPa^{-1}$ (Devi et al. 2010).

In a study of nine southern US commercial cultivars, eight cultivars showed TR_{lim} above a VPD threshold ranging from 1.81 to 2.25 kPa, and TR rate increase above the threshold ranged from −10.10 to 3.63 mg $H_2O\,m^{-2}\,s^{-1}$ (Devi and Sinclair 2011). In a controlled-environment study with six peanut genotypes with different breeding backgrounds, three genotypes expressed TR_{lim} trait with VPD breakpoint range of 2.3–2.8 kPa. These genotypes had decreasing TR above their VPD breakpoint (Shekoofa et al. 2015).

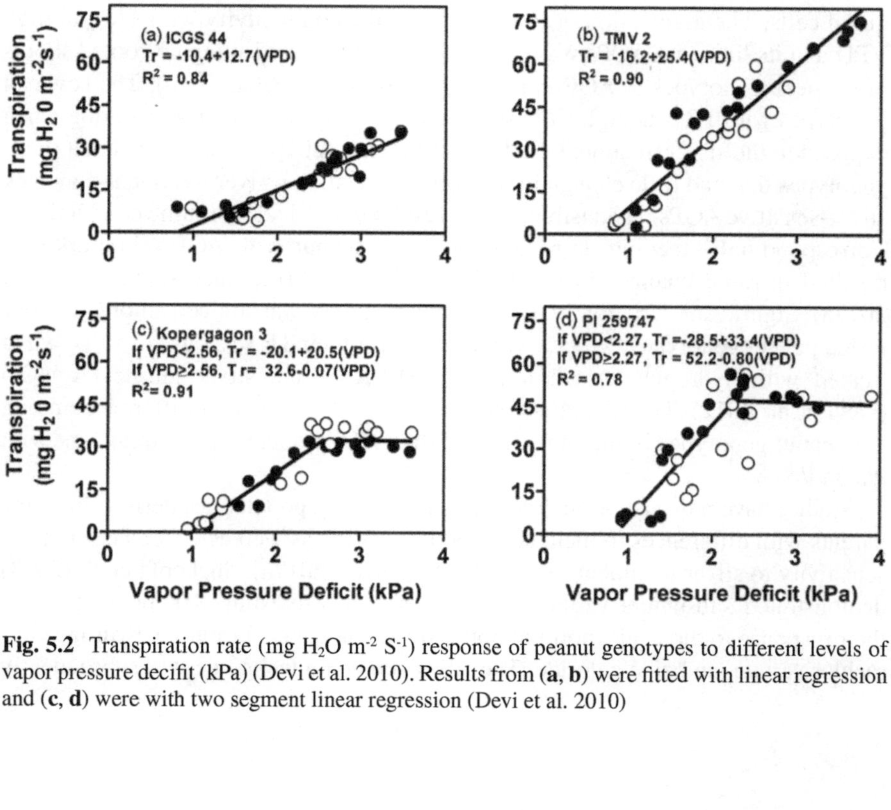

Fig. 5.2 Transpiration rate (mg $H_2O\ m^{-2}\ S^{-1}$) response of peanut genotypes to different levels of vapor pressure decifit (kPa) (Devi et al. 2010). Results from (**a, b**) were fitted with linear regression and (**c, d**) were with two segment linear regression (Devi et al. 2010)

A study of TR response in a controlled-environment VPD study with six parents of existing peanut RIL population showed expression of the TR_{lim} trait with decreases in TR above VPD breakpoints ranging from 1.61 to 2.90 kPa (Shekoofa et al. 2013). In another study with breeding lines (Shekoofa et al. 2017a), 10 lines out of 15 runner-type and 4 out of 12 Virginia-type lines were found to express the TR_{lim} trait. In runner-type and Virginia-type peanut, the range of VPD breakpoint for TR_{lim} trait was found to be between 2.16 and 3.38 kPa, and 1.90 and 2.52, respectively. In these RILs the TR increase above the VPD breakpoint was very small or negative in these lines.

In another field study with 18 peanut genotypes under soil-drying conditions and high VPD, a negative relation between VPD and stomatal conductance was observed (Balota et al. 2012). The decrease in stomatal conductance at high VPD is indicative of expression of the TR_{lim} trait under these conditions.

5.2.2 *Hydraulic Conductivity and Aquaporins*

Genotypes expressing TR_{lim} with increasing atmospheric VPD were found in species other than peanut to have low hydraulic conductivity in their leaves (Sinclair et al. 2008; Choudhary et al. 2013, 2014). It was hypothesized that the low hydraulic conductance in the leaves may reflect a population of water-transport-mediating proteins, that is, aquaporins, that result in a poorer ability to transport water to guard cells. The association of low leaf hydraulic conductivity with TR_{lim} at high VPD and its link with AQPs was originally studied by subjecting de-rooted shoots of soybean genotypes to AQP inhibitors (Sadok and Sinclair 2010). The soybean genotypes with TR_{lim} at high VPD showed the least decrease in transpiration when exposed to the silver treatment. Sadok and Sinclair (2010) hypothesized that those genotypes that had little change in transpiration with the silver treatment may lack silver-sensitive AQPs responsible, which results in low leaf hydraulic conductivity. Correspondingly, feeding de-rooted shoots of peanut with AQP inhibitors also resulted in rapid changes in plant TR (Devi et al. 2012; Shekoofa et al. 2013, 2017a). Significant variation in transpiration response was noticed among the four tested peanut genotypes, which were different for their TR response to VPD, when treated with hydrogen tetrachloroaurate ($HAuCl_4$) and silver nitrate ($AgNO_3$) (Devi et al. 2012). These results were interpreted to indicate a difference among the peanut genotypes in the sensitivity of their AQP populations to treatment with Au and/or Ag.

Studies have now also been done using peanut RIL populations derived from the parents with differences in their TR response to VPD as well as differences in their sensitivity to silver treatment (Shekoofa et al. 2013, 2017a). Shekoofa et al. (2013) demonstrated substantial variation among parents of existing RIL population for their response to three aquaporin inhibitor treatments of silver nitrate, hydrogen tetrachloroaurate, and zinc chloride. The silver screening technique was subsequently used as an alternate screen in identifying the TR_{lim} trait in two peanut populations (Shekoofa et al. 2017a).

The abundance of AQP transcripts in peanut genotypes differing in the expression of TR_{lim} trait was established in a recent study (Devi et al. 2016). Peanut cultivars with the TR_{lim} trait had decreased AQP transcript abundance for the four AQPs tested out of six when exposed to high VPD compared to low VPD environment. The genotypes with no TR_{lim} trait increased their AQP transcript abundance with an increase in VPD. In the same study, AQP inhibitor treatment resulted in increased abundance of AQP transcripts in both the TR_{lim} and no TR_{lim} genotypes, possibly to compensate blockage.

5.2.3 Response to the Environment

A concern in exploring genotypes for the TR_{lim} trait is the potentially confounding effects of environmental conditions and stage of plant development in its expression. Specifically, for genotype comparisons of TR_{lim} to be relevant for crop production, a key issue is whether temperature influences expression of the trait such that selections have to be done under controlled temperature conditions (Shekoofa et al. 2015).

In a recent study among peanut genotypes, there was a difference in expression of the TR_{lim} trait between controlled-environment experiments done at 31 and at 36 °C (Shekoofa et al. 2015). In the experiment conducted at 31 °C, three out of six peanut genotypes (N05008, Georgia Green, and HTS 02-05) expressed the TR_{lim} trait under high VPD, while the other three genotypes expressed a linear response in TR to increasing VPD (Table 5.1). In the second experiment under 36 °C, all six genotypes expressed a linear response in TR to increasing VPD (Table 5.1). None of the six genotypes were found to express the TR_{lim} trait in the field where the plants were subjected to high temperature (Shekoofa et al. 2015).

5.3 Inheritance of TR_{lim} Trait

Two genetic populations of peanut were studied for the expression of TR_{lim} trait by Shekoofa et al. (2013). Each of the populations was derived from parents that differed in their TR response to VPD. In these studies, screening with silver, an AQP inhibitor, was employed as an alternative method to the various laborious and impractical approach of direct measure for TR_{lim} response. A large range of response in TR to silver treatment was observed within each population with a substantial fraction of progeny lines expressing more sensitivity to the silver test, that is, loss of TR_{lim}, than either parent. These results indicated dominance for a genetic loss in the TR_{lim} trait in this population.

Table 5.1 Regression results of transpiration rate vs. vapor pressure deficit for chamber studies (31/26 and 36/26 °C day/night temperature) (Shekoofa et al. 2015)

32/26 °C						36/26 °C		
Genotypes	*n*	Slope1 ± S.E.	Breakpoint (X_0) ± S.E. (kPa)	Slope2 ± S.E.	R^2	Slope1 ± S.E.	Breakpoint (X_0) ± S.E. (kPa)	R^2
SPT0607	24	13.7 ± 1.26	Linear	–	0.85	30.8 ± 3.45	Linear	0.78
HTS02-05	24	12.2 ± 1.99	2.8 ± 0.22	−10.7 ± 8.30	0.72	28.5 ± 4.77	Linear	0.61
Bailey	24	11.0 ± 1.35	Linear	–	0.75	29.8 ± 4.68	Linear	0.64
Georgia Green	24	14.0 ± 2.73	2.3 ± 0.24	6.3 ± 4.77	0.70	21.81 ± 3.55	Linear	0.63
N05006	24	12.1 ± 1.63	Linear	–	0.72	15.42 ± 3.58	Linear	0.50
N05008	24	16.5 ± 3.19	2.6 ± 0.19	17.1 ± 13.23	0.70	25.98 ± 2.81	Linear	0.79

Those results represented by the two-segment linear regression include results for the two slopes and the VPD at the breakpoint

Recently, Shekoofa et al. (2017b) studied the expression of the TR_{lim} trait in two peanut progeny RIL populations: runner and virginia type. Two levels of screening were performed on both populations. Initially the populations were screened by feeding silver ions to a large number of genotypes in each population. Then, a smaller number of RILs were selected for direct measure of TR response to VPD. The study showed that an effective expression of the TR_{lim} trait occurred in about 25% of the progeny in each of the two populations indicating the possibility of transferring the TR_{lim} trait to progeny genotypes.

References

Balota M, McGrath S, Isleib TG et al (2012) Transpiration response to vapor pressure deficit in field grown peanut. Peanut Sci 39(1):53–61

Bhatnagar-Mathur P, Devi MJ, Reddy SD et al (2007) Stress-inducible expression of At DREB1A in transgenic peanut (*Arachis hypogea* L.) increases transpiration efficiency under water-limiting conditions. Plant Cell Rep 26:2071–2082

Choudhary S, Sinclair TR, Prasad PV (2013) Hydraulic conductance of intact plants of two contrasting sorghum lines, SC15 and SC1205. Funct Plant Biol 40(7):730–738

Choudhary S, Sinclair TR, Messina CD et al (2014) Hydraulic conductance of maize hybrids differing in transpiration response to vapor pressure deficit. Crop Sci 54(3):1147–1152

Devi MJ, Sinclair TR (2011) Diversity in drought traits among commercial southeastern US peanut cultivars. Int J Agron. http://dx.doi.org/10.1155/2011/754658

Devi MJ, Sinclair TR, Vadez V et al (2009) Peanut genotypic variation in transpiration efficiency and decreased transpiration during progressive soil drying. Field Crop Res 114(2):280–285

Devi MJ, Sinclair TR, Vadez V (2010) Genotypic variation in peanut for transpiration response to vapor pressure deficit. Crop Sci 50(1):191–196

Devi MJ, Sadok W, Sinclair TR (2012) Transpiration response of de-rooted peanut plants to aquaporin inhibitors. Environ Exp Bot 78:167–172

Devi MJ, Sinclair TR, Jain M et al (2016) Leaf aquaporin transcript abundance in peanut genotypes diverging in expression of the limited-transpiration trait when subjected to differing vapor pressure deficits and aquaporin inhibitors. Physiol Plant 156(4):387–396

Sadok W, Sinclair TR (2010) Transpiration response of 'slow-wilting'and commercial soybean (*Glycine max* (L.) Merr.) genotypes to three aquaporin inhibitors. J Exp Bot 61(3):821–829

Sharma KK, Lavanya M (2002) Recent developments in transgenics for abiotic stress in legumes of the semi-arid tropics. JIRCAS Work Rep 23:61–73

Shekoofa A, Devi JM, Sinclair TR et al (2013) Divergence in drought-resistance traits among parents of recombinant peanut inbred lines. Crop Sci 53(6):2569–2576

Shekoofa A, Rosas-Anderson P, Sinclair TR et al (2015) Measurement of limited-transpiration trait under high vapor pressure deficit for peanut in chambers and in field. Agron J 107(3): 1019–1024

Shekoofa A, Sinclair TR, Annubon C et al (2017a) Expression of the limited-transpiration trait under high vapor pressure deficit in peanut RIL populations: runner and Virginia types. J Agr Crop Sci. doi:10.1111/jac.12204

Shekoofa A, Sinclair TR, Balota M et al (2017b) Drought tolerant traits in VA type peanuts: field and growth chamber studies (in preparation)

Sinclair TR, Vadez V (2012) The future of grain legumes in cropping systems. Crop Pasture Sci 63:501–512. doi:10.1071/ CP12128

Sinclair TR, Zwieniecki MA, Holbrook NM (2008) Low leaf hydraulic conductance associated with drought tolerance in soybean. Physiol Plant 132(4):446–451

Chapter 6
Chickpea

Kaliamoorthy Sivasakthi, Mainassara Zaman-Allah, Murugesan Tharanya, Jana Kholová, Thiyagarajan Thirunalasundari, and Vincent Vadez

Chickpea (*Cicer arietinum* L.) is the second most important pulse crop worldwide next to beans. It is a cool season annual crop that belongs to the Leguminosae family. It is a rich source of protein (20–30%), carbohydrates (~40%), and minerals and thus serves as an important source of nutrients in vegetarian diets, especially in developing countries (Jukanti et al. 2012). The average production of chickpea was 14.2 million tons from an area of 14.8 million ha giving an average yield of 0.96 t ha^{-1} (FAO 2014). About 90% of this crop was grown under residual soil moisture in the arid and semiarid regions of the world (Kumar and Abbo 2001). Terminal drought is one of the major constraints limiting chickpea productivity (Krishnamurthy et al. 2010) that causes substantial annual yield losses up to 50% (Ahmad et al. 2005). Plant traits that allow soil water conservation (limited or restricting TR) under well-watered conditions during the vegetative stage may allow greater water availability for successful grain filling during terminal drought.

K. Sivasakthi • M. Tharanya
Bharathidasan University, Tiruchirappalli, Tamil Nadu, India

International Crops Research Institute for the Semi-Arid Tropics (ICRISAT), Greater Hyderabad, Telangana, India

M. Zaman-Allah
International Center for Maize and Wheat Improvement (CIMMYT), P.O Box, MP 163, Mount Pleasant, Harare, Zimbabwe

J. Kholová • V. Vadez (✉)
International Crops Research Institute for the Semi-Arid Tropics (ICRISAT), Greater Hyderabad, Telangana, India
e-mail: v.vadez@cgiar.org

T. Thirunalasundari
Bharathidasan University, Tiruchirappalli 620 024, Tamil Nadu, India

T.R. Sinclair (ed.), *Water-Conservation Traits to Increase Crop Yields in Water-deficit Environments*, SpringerBriefs in Environmental Science,
DOI 10.1007/978-3-319-56321-3_6

6.1 Transpiration Response to Soil Drying

A decline in transpiration rate at high soil moisture was hypothesized to increase chickpea production for cases dependent on stored soil moisture. Soltani et al. (2000) examined this possibility by using a predecessor of the simple simulation model (SSM) to simulate chickpea yield in Iran at Tabriz (38°8'N, 42°17'E and 1364 m a.s.l.) and Gorgan (36°51'N, 54°16'E and 13 m a.s.l.). In this model, sensitivity to soil drying is simulated by defining a soil water threshold at which transpiration and leaf expansion begin to decline. The baseline thresholds they used in the simulations were 0.34 FTSW for transpiration and 0.48 FTSW for leaf expansion based on pot experiments. Simulations were done by increasing and decreasing the thresholds up to 40%. At Gorgan, yield was changed little, altering the two thresholds. At Tabriz, increasing the transpiration threshold resulted in small increases in yield. The reason for the relatively small impact on yield may be a result of chickpea in these simulations being grown under residual moisture. Several intermittent events of water-deficit stress may have resulted in a larger impact of this trait on yield.

There now have been several experimental studies undertaken to identify genotypes that might have especially high thresholds for the initiation in the decrease of transpiration with soil drying (Table 6.1). Zaman-Allah et al. (2011a) investigated eight chickpea genotypes, which were selected for having comparable phenology but contrasted in seed yield under terminal water deficit in the field. Upon exposure to progressively greater water deficit in outdoor conditions, transpiration rate of those genotypes originally identified as drought tolerant had initiation of partial stomatal closure at higher FTSW (~0.40) than the FTSW threshold for the water-deficit-sensitive genotypes (~0.30). However, the growth of the tolerant genotypes was less than the sensitive genotypes likely reflecting the greater degree of stomatal closure during the experiment.

In a companion experiment performed under controlled greenhouse conditions, Zaman-Allah et al. (2011a) found with the same genotypes no significant difference in FTSW threshold values between water-deficit-tolerant genotypes and water-deficit-sensitive ones. Even though there was no difference in expression of the FTSW threshold, the growth of the water-deficit-sensitive genotypes was greater than the water-deficit-tolerant lines. Zaman-Allah et al. (2011a) hypothesized that the lack of difference among genotypes in the FTSW threshold in the greenhouse experiment may have been due to the lower vapour pressure deficits (VPD) in the greenhouse experiment as compared to the outdoor experiment. Under the lower VPD conditions in the greenhouse, the sensitivity to soil drying may have been decreased so that discrimination among genotypes was not apparent. In the field, with higher VPD sensitivity to VPD may have been the key factor resulting in discrimination between sensitive and tolerant genotypes.

Sivasakthi et al. (manuscript under communication) compared a pair of chickpea genotypes contrasting in shoot growth rate and root architecture. There was a significant difference in the FTSW thresholds between low (0.61) and high (0.52) shoot growth genotypes. This result confirmed those of Zaman-Allah with a larger range of genotypes (Zaman-Allah et al. 2011a) and where lower growth genotypes also had a higher FTSW threshold.

Table 6.1 Description of conditions for experiments measuring transpiration response to soil drying (dry down) including growth stage, place, and vapor pressure deficit (VPD)

	Growth stage	Growth place	Mean VPD range of both growing and measurement period (k Pa)	No. of genotypes	Discrimination of FTSW threshold between genotypes
Zaman-Allah et al. (2011a)	Vegetative	Glasshouse	0.63–2.85	8	No
Zaman-Allah et al. (2011a)	Vegetative	Outdoor	0.18–2.80	8	Yes
Anbazhagan et al. (2015)	Vegetative	Glasshouse	0.5–3.0	4	Yes
Sivasakthi et al. (Manuscript under communication)	Vegetative	Glasshouse	0.60–2.80	2	Yes
Pushpavalli et al. (2014)	Reproductive	Glasshouse	0.25–2.75	10	No
Pang et al. (2016)	Reproductive	Glasshouse	0.12–1.50	2	No

Pushpavalli et al. (2014) investigated transpiration response of ten chickpea genotypes to soil drying under controlled environment. These genotypes were selected based on differences in seed yield under terminal water deficit in the field. In this study, progressive water-deficit stress was initiated at the onset of flowering, i.e., later than in the above-reported trials. The results from this experiment showed slightly higher FTSW threshold values than vegetative phase FTSW values, and FTSW threshold values did not discriminate water-deficit tolerant from sensitive genotypes. Pang et al. (2016) investigated two chickpea genotypes with similar phenology but contrasting for seed yield under field conditions. They showed that there were no differences in the FTSW threshold values between the two chickpea genotypes in the reproductive phase. Both these studies indicated lower genotypic variation when the crop is tested during reproductive stages.

Anbazhagan et al. (2015) examined a possible gene basis for the response of chickpea genotypes to soil drying. A study was done on chickpea transgenic events harboring rd29A:DREB1A transcription factor (under the control of stress-inducible rd29A promoter), all with a single copy of the gene. Their results showed that transgenic plants had lower FTSW thresholds for initiation of transpiration decline with soil drying than untransformed plants.

6.2 Transpiration Response to Increasing Vapor Pressure Deficit

Given that VPD appears to be an important confounding factor in interpreting results from soil drying experiments, direct responses of transpiration rate to atmospheric VPD needed to be documented. Zaman-Allah et al. (2011a) examined in

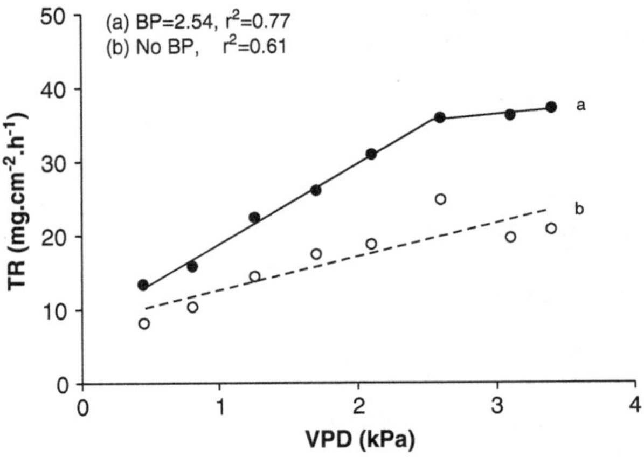

Fig. 6.1 Transpiration rate (TR, mg $H_2O.cm^{-2}\ h^{-1}$) under well-watered conditions of sensitive (**a**) Line ICC 4814 and tolerant (**b**) Line ICC 14799 chickpea genotypes exposed to increasing VPD regimes under controlled conditions of a growth chamber. Plants were assessed at the vegetative stage and each point represents the mean of four replicates (Zaman-Allah et al. 2011a)

detail the transpiration response to increasing VPD of eight chickpea genotypes differing in their drought sensitivity, as discussed previously. In the outdoor experiment, water-deficit-sensitive genotypes had a higher TR than water-deficit-adapted genotypes in the middle part of the day (corresponding to the highest VPD).

The controlled-environment experiment allowed full documentation of the TR response over a wide range of VPD. These results confirmed that the water-deficit-sensitive genotypes had greater increases in TR response with increasing VPD than the water-deficit-adapted genotypes. However, half of the water-deficit-sensitive genotypes expressed a VPD breakpoint in TR (Fig. 6.1).

These previous results were also in agreement with lysimetric water extraction data (Zaman-Allah et al. (2011b), where water-deficit-adapted genotypes (e.g., ICC 867 and ICC 14799) had lower water extraction at 28–38 DAS than water-deficit-sensitive genotypes (e.g., ICC 8058 and ICC 4814). The water extraction data correlated with lower stomatal conductance of ICC 867 and ICC 14799. At 66 DAS (pod filling stage), another similar set of plants were used to measure transpiration rate under naturally changing atmospheric VPD outdoors, and it showed that transpiration rate response was opposite to vegetative stage assessment. That is, the water-deficit-adapted genotypes (e.g., ICC 867 and ICC 14799) had higher transpiration rate than water-deficit-sensitive genotypes (e.g., ICC 4814 and ICC 8058; Zaman-Allah et al. 2011a).

These results highlighted distinct differences in the pattern of water use differed between water-deficit-adapted and water-deficit-sensitive lines during vegetative and reproductive stage. The water-deficit-adapted genotypes had lower canopy conductance at vegetative stage by slower canopy development; by contrast, maximized canopy development at reproductive stage resulted in higher canopy conductance than water-deficit-sensitive lines. Plant growth rate and VPD might be critical factor for changes in water uptake during different growth stages.

Anbazhagan et al. (2015) examined the transpiration rate response of transgenic chickpea plants with transformations harboring rd29A:DREB1A transcription factor (under the control of stress-inducible rd29A promoter), all with a single copy of the gene. The TR in response to VPD of transformed plants was compared to untransformed control plants with similar plant leaf area groups. All measured transgenic events were associated with lower TR slopes than untransformed control plants under increasing VPD.

Sivasakthi et al. (manuscript under communication) compared four chickpea genotypes contrasting in plant growth rate and root architecture and found that genotypes with higher growth rate late in the season had a linearly increasing TR throughout the whole range of tested VPD. By contrast genotypes with high early growth rate had lower TR and also had TR restriction at VPD level above 2 kPa, and then decreasing transpiration rate at VPD greater than the VPD breakpoint. VPD breakpoints ranged from 2.5 to 2.7 kPa in genotypes with greater early growth rate. Another similar study between mapping population parents, contrasting for plant growth rate and root parameters and their recombinant inbred line progenies, showed that TR response to VPD of the parents and progenies was described by the two-linear-segment regression model. The higher early growth parent and progeny genotypes had lower TR with lower VPD breakpoints (2.4 and 1.8 kPa) than the breakpoint of the higher late growth parent and progeny genotypes (3.1 kPa). Also, at VPD greater than the VPD breakpoints TR gradually decreased in the early growth parent and progeny genotypes than late growth parent and progeny genotypes (Sivasakthi et al. manuscript under communication).

Canopy temperature measurements can be used as a proxy for stomatal conductance (Jones 1999; Jones et al. 2002). Canopy temperature was then assessed at both vegetative and early podding stages using an infrared thermal camera and showed that water-deficit-adapted genotypes had a higher canopy temperature than water-deficit-sensitive genotypes under well-water conditions at vegetative stage and in the middle part of the day (Zaman-Allah et al. 2011a, b). It was interpreted that water-deficit-adapted genotypes restricted the transpiration through partial stomatal closure under high VPD conditions which resulted in an increased canopy temperature, even in the absence of a soil water deficit.

6.3 Hydraulic Limitation

Recent research indicates that the rapid response of the transpiration upon a VPD increase is driven by hydraulic signals rather than biochemical ones (i.e., ABA) (Comstock 2000). Recent investigations of chickpea genotypes attempted to locate the morphological source of a possible hydraulic limitation in these plants (Sivasakthi et al. manuscript under communication). This study involved a comparison of the transpiration response to increasing VPD in whole plants and de-rooted shoots. Results showed that de-rooted shoots had higher TR than whole plants, especially under high VPD conditions. More interestingly, de-rooted shoot had no transpiration restriction under high VPD conditions. By contrast, whole plants expressed transpiration restriction at VPD above 2.5–3.0 kPa. The authors hypothesized that the roots were likely to be a major source of hydraulic limitation in chickpea.

Radial water flow in the roots is driven by pressure gradients (hydrostatic & osmotic) encountering a major hydraulic resistance at the endodermis where the apoplastic water movement is limited by Casparian bands. It has been suggested that apoplastic pathways predominate in water transport in crop species like maize (Zimmermann and Steudle 1998). To investigate the role of apoplastic-mediated water flow in chickpea, a perfusion technique was applied to root segments causing at least a partial blockage of the apoplastic compartment (Ranathunge et al. 2005). For this experiment, 25-day-old, hydroponically grown plants were treated with 1 mM $K_4[Fe(CN)_6]$ and 0.5 mM $CuSO_4$ to cause the formation of rusty brown crystals (precipitates) of $Cu_2[Fe(CN)_6]_6$ or $Cu[CuFe(CN)_6]$ in the apoplast. Root anatomy observations in chickpea following this technique showed clear insoluble crystals in apoplastic pores or pathways (Sivasakthi et al. manuscript under submission). VPD-sensitive genotypes had higher apoplastic inhibition than VPD-insensitive genotypes. This response was contrary to AQP-mediated, water transport pathway inhibition. The level of inhibition differed between apoplast and AQP-mediated, water transport with the apoplastic blockage resulting in more decline in transpiration than the inhibition of AQP-mediated pathway.

Recently, hydroponically grown plants (26 days old) were used to measure root hydraulic conductivity of chickpea (Sivasakthi et al. manuscript under communication). The measurements were done on de-topped roots bathing in water that were carefully placed in a pressure chamber (PMS instruments, Corvallis, Oregon, USA). The stem was sealed in the pressure chamber using silicon glue and polyvinyl siloxane (Coltene President Company, Switzerland) to prevent pressure leakage. The pressure was applied in the root medium at the range from 0.1 to 0.3 MPa. The root exudate (xylem sap) was collected at each pressure level over three 5 min intervals using Eppendorf cones filled with tissue paper (Kimtech Science, Ontario, USA). The root hydraulic conductivity was calculated by normalizing the amount of exudate by the root surface area and pressure. The root surface area was estimated by scanning with Shimadzu scanner and analyzing with Winrhizo software (Winrhizo, Regent Ltd., Canada).Under non-treated conditions (without any inhibitions), a VPD-insensitive genotype with high TR had a higher root conductivity than VPD-sensitive with low TR genotypes. After AQP inhibition (20 μM $HgCl_2$), root hydraulic conductivity decreased in all tested genotypes. The VPD-insensitive genotypes had a higher decrease in root hydraulic conductivity than VPD-sensitive genotypes. When the apoplastic pathway with precipitants as described above, root hydraulic conductivity was also highly decreased in all tested genotypes with VPD-insensitive genotypes having a higher decrease than VPD-sensitive genotypes.

6.4 Aquaporin Mediated Water Transport Pathway

The water transport pathways in both roots and leaves include water transport across cell membranes. This cell-to-cell water pathway across membranes is facilitated diffusion membrane-embedded proteins identified as aquaporins (AQP). Hence, AQPs can play a key role in regulating hydraulic conductivity (Chaumont and Tyerman 2014; Maurel et al. 2015). One approach in testing AQP involvement in determining hydraulic conductivity is a pharmacological approach of feeding plants AQP inhibitors.

Sivasakthi et al. (manuscript under communication) investigated the possible role of AQP in regulating water transport in chickpea by applying inhibitors $HgCl_2$ and H_2O_2. Four chickpea genotypes contrasting in TR response to VPD were grown hydroponically and at 26 days after sowing subjected to the AQP inhibitors (100 & 200 μM $HgCl_2$ and 1 mM H_2O_2). Within 30 min after feeding intact plants the AQP inhibitors, VPD-insensitive genotypes showed a greater decline in TR than VPD-sensitive genotypes (Fig. 6.2). Soil grown, de-rooted shoot fed with AQP inhibitor showed responses similar to those observed with whole plant treatments. Therefore, these AQP inhibition studies indicate that there may be differences in AQPs among genotypes in at least the shoots.

To extend the study of the influence of AQP, 45 AQPs were identified in chickpea and characterized in different tissue under abiotic stress conditions (Reddy et al.

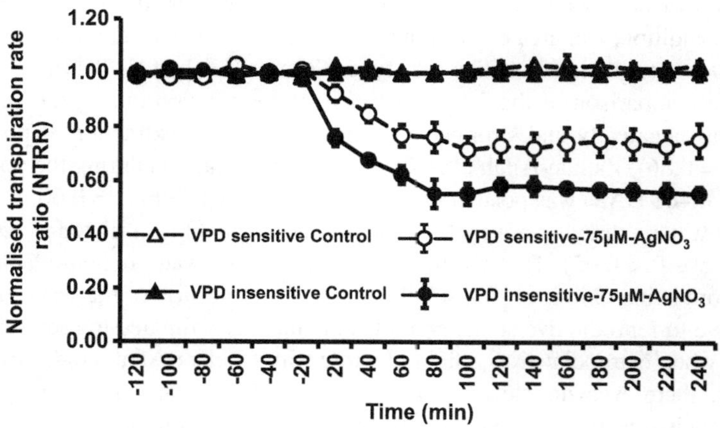

Fig. 6.2 Decline in transpiration of two chickpea genotypes (VPD sensitive and insensitive) contrasting for transpiration response to VPD. Plants were grown in a hydroponics system and tested at vegetative stage (26 days old) in high VPD (3.1 kPa) conditions. The aquaporin inhibitor (75 μM-$AgNO_3$) was fed after 120 min transpiration measurement. The *solid lines with closed circles* represent VPD-insensitive inhibited genotype and *dotted lines with open circles* represent VPD-sensitive inhibited genotype. The *solid lines with closed triangles* represent VPD-insensitive non-inhibited genotype and *dotted lines with open triangles* represent VPD-sensitive non-inhibited genotype. Each treatment data point represents the NTRR means (±SE) of six replicates per genotype

manuscript under communication). Based on these results, 13 AQPs (7PIPs, 4 TIPs and 2 NIPs) were selected as candidates for involvement in response to VPD treatments in different chickpea genotypes. AQP gene expression analysis revealed that overall, lower-TR genotypes had a lower AQP gene expression than a higher-TR genotype under high VPD (4.21 kPa) conditions.

6.5 Water Extraction During Growing Season

Water conservation has been assessed in a field-like situation by using a lysimetric system [PVC tubes filled with Vertisol, which provide space and soil volume for water exploration similar to the field conditions]. The continuous weighing of these tubes resulted in a transpiration profile over the entire cropping cycle (Vadez et al. 2008, 2013, 2014; Ratnakumar et al. 2009). Zaman-Allah et al. (2011b) used this system to investigate water use by 20 chickpea genotypes, which had the same phenological characteristics but contrasting seed yield under terminal water deficit in field trials. Water-deficit-adapted genotypes extracted less water than water-deficit-sensitive genotype under well-watered conditions (Fig. 6.3a), whereas under water-deficit conditions, it extracted more water than the water-deficit-sensitive genotype (Fig. 6.3b). Also, the water-deficit-adapted genotypes had lower amounts of water uptake than water-deficit-sensitive genotypes at vegetative stages. Interestingly, water-deficit-adapted genotypes had lower shoot mass than water-deficit-sensitive genotypes in both well-water and water-deficit conditions during the vegetative stage. In addition, leaf area expansion rate was lower in water-deficit-adapted genotypes than water-deficit-sensitive genotypes (Zaman-Allah et al. 2011a).

In the comparison of the 20 genotypes, water extraction during the vegetative stage was negatively and strongly correlated with water extraction at pod filling stage ($r = 0.86$) (Zaman-Allah et al. 2011b). Water extraction during the pod filling period (48–61 DAS) was positively correlated with seed yield ($r = 0.66$, $P < 0.05$), whereas water extraction at vegetative stage was negatively correlated to seed yield ($r = -0.50$, $P < 0.05$). That is, the yield advantage of water-deficit-adapted lines came from being able to shift a small but significant portion of the water use from vegetative to reproductive stage, and this amount was estimated to represent about 25 mm when extrapolated to a field plant density (Zaman-Allah et al. 2011b). By contrast, there was no clear discrimination between water-deficit-adapted and water-deficit-sensitive genotypes on the basis of total water used, rooting depth, and root length density measured during the reproductive stage. Overall, this study showed that terminal water-deficit adaptation in chickpea mostly depended on having water available for the grain filling period, which in this case was due to a more conservative use of water in the vegetative stage of cropping cycle.

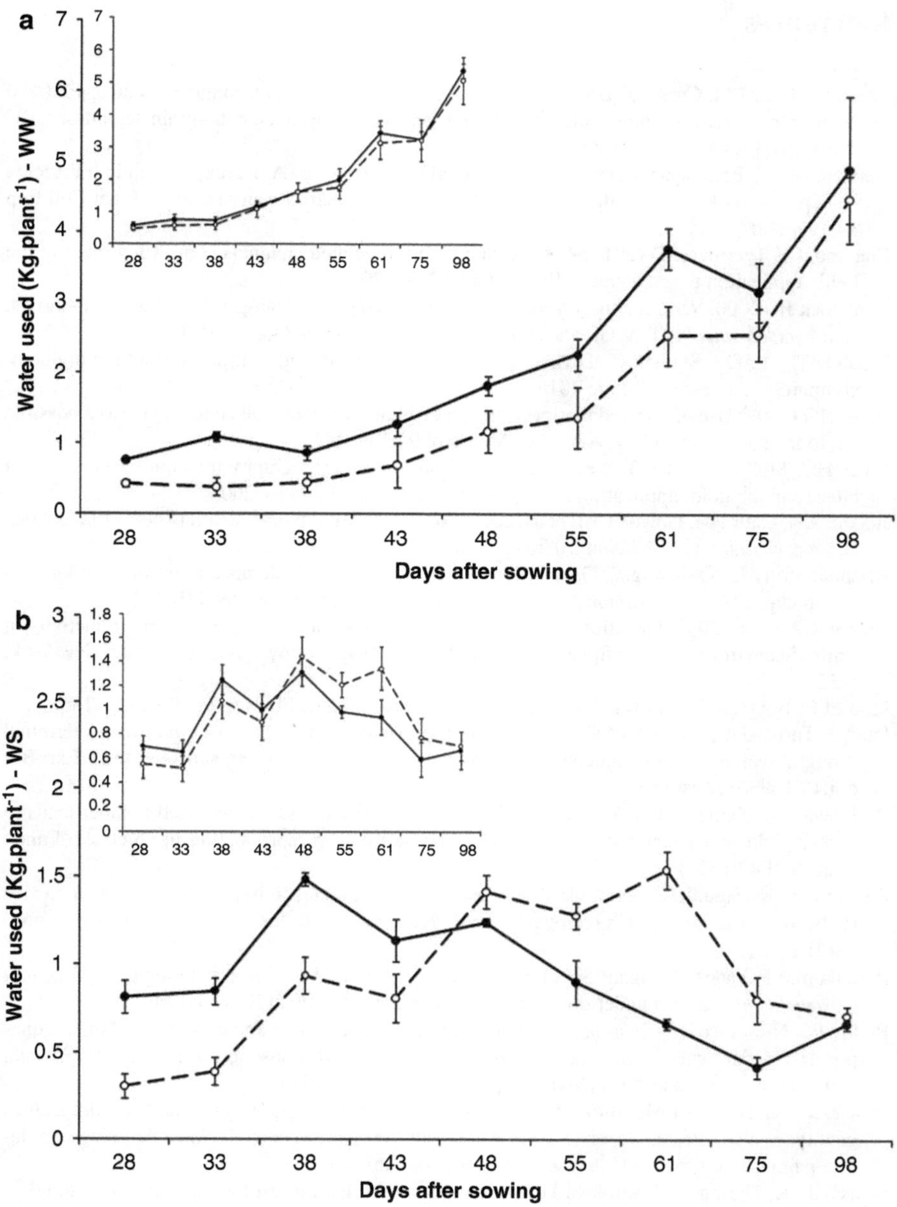

Fig. 6.3 Variation in water uptake profile in two chickpea genotypes contrasting for the drought stress index (DTI) under terminal drought stress (sensitive ICC 8058, *closed symbols and solid lines*; tolerant ICC 867, *open symbols and dashed lines*) and grown under (**a**) well-watered and (**b**) water-stressed conditions (Zaman-Allah et al. 2011b)

References

Ahmad F, Gaur PM, Croser J (2005) Chickpea *(Cicer arietinum L.)* In: Singh RJ, Jauhar PP (eds) Genetic resources, chromosome engineering, and crop improvement—grain legumes. CRC Press, Boca Raton

Anbazhagan K, Bhatnagar-Mathur P, Vadez V et al (2015) DREB1A overexpression in transgenic chickpea alters key traits influencing plant water budget across water regimes. Plant Cell Rep 34:199–210

Chaumont F, Tyerman SD (2014) Aquaporins: highly regulated channels controlling plant in the field: application to grapevine. J Exp Bot 53:2249–2260

Comstock JP (2000) Variation in hydraulic architecture and gas-exchange in two desert sub-shrubs, *Hymenoclea salsola* (T. & G.) and *Ambrosia dumosa* (Payne). Oecol 125(1):1–10

FAOSTAT, FAO Statistics Division (2014) Available at: http://faostat3.fao.org/com/compare/E. Accessed 27 June 2016

Jones HG (1999) Use of infrared thermometry for estimation of stomatal conductance as a possible aid to irrigation scheduling. Agric For Meteorol 95:139–149

Jones HG, Stoll M, Santos T et al (2002) Use of infrared thermography for monitoring stomatal closure in the field: application to grapevine. J Exp Bot 53:2249–2260

Jukanti AK, Gaur PM, Gowda CLL et al (2012) Nutritional quality and health benefits of chickpea (*Cicer arietinum* L.) Br J Nutr 108:S11–S26

Krishnamurthy L, Kashiwagi J, Gaur PM et al (2010) Sources of tolerance to terminal drought in the chickpea (*Cicer arietinum L.*) minicore germplasm. Field Crop Res 119:322–330

Kumar J, Abbo S (2001) Genetics of flowering time in chickpea and its bearing on productivity in semiarid environments. In: Spaks DL (ed) Advances in agronomy, vol 2. Academic, New York, p 22

Maurel C, Boursiac Y, Luu DT et al (2015) Aquaporins in plants. Physiol Rev 95:1321–1358

Pang J, Turner NC, Khan T et al (2016) Response of chickpea (*Cicer arietinum* L.) to terminal drought: leaf stomatal conductance, pod Abscisic acid concentration, and seed set. J Exp Bot doi:10.1093/jxb/erw153

Pushpavalli R, Zaman-Allah M, Turner NC et al (2014) Higher flower and seed number leads to higher yield under water stress conditions imposed during reproduction in chickpea. Funct. Plant Biol 42:162–174

Ranathunge K, Steudle E, Lafitte R (2005) Blockage of apoplastic bypass-flow of water in rice roots by insoluble salt precipitates analogous to a Pfeffer cell. Plant Cell Environ 28(2):121–133

Ratnakumar P, Vadez V, Nigam SN et al (2009) Assessment of transpiration efficiency in peanut (*Arachis hypogaea* L.) under drought by lysimetric system. Plant Biol 11:124–130

Reddy PS, Nitnavare RB, Sivasakthi K et al Genome wide screening and characterization of aquaporins (AQPs) genes in chickpea (*Cicer arietinum* L.) in response to different developmental and abiotic stress conditions (Manuscript under communication)

Sivasakthi K, Tharanya M, Kholová J et al Difference in plant vigour and canopy conductance genotypes also differed dependence of water transport pathways in chickpea mapping population parents and progeny (Manuscript under communication)

Sivasakthi K, Tharanya M, Kholová J et al Genetic variability of transpiration response to soil drying in contrasting chickpea genotypes (Manuscript under communication)

Sivasakthi K, Tharanya M, Zaman-Allah M et al Differences in hydraulic conductivity of chickpea roots (*Cicer arietinum* L.) relate to crop adaptation to drought environments. (Manuscript under communication)

Soltani A, Khooie FR, Ghassemi-Golezani K et al (2000) Thresholds for chickpea leaf expansion and transpiration response to soil water deficit. Field Crop Res. 68:205–210

Vadez V, Rao S, Kholova J et al (2008) Root research for drought tolerance in legumes: quo vadis? J Food Legumes 21:77–85

Vadez V, Kholová J, Yadav RS et al (2013) Small temporal differences in water uptake among varieties of pearl millet [*Pennisetum glaucum* (L.) R. Br.] are critical for grain yield under terminal drought. Plant Soil 371:447–462

Vadez V, Kholova J, Medina S et al (2014) Transpiration efficiency: new insights into an old story. J Exp Bot 65(21):6141–6153. doi:10.1093/jxb/eru040

Zaman-Allah M, Jenkinson DM, Vadez V (2011a) Chickpea genotypes contrasting for seed yield under terminal drought stress in the field differ for traits related to the control of water use. Funct Plant Biol 38(4):270–281

Zaman-allah M, Jenkinson DM, Vadez V (2011b) A conservative pattern of water use, rather than deep or profuse rooting, is a critical for the terminal drought tolerance of chickpea. J Exp Bot 62(12):4239–4252

Zimmermann HM, Steudle E (1998) Apoplastic transport across young maize roots: effect of the exodermis. Planta 206:7–19

Chapter 7
Lentil

Michel Edmond Ghanem, Hélène Marrou, Julie Guiguitant, and Fatima ez-zahra Kibbou

Lentil (*Lens culinaris* Medikus) is an important cool season legume, particularly within countries in North America, South Asia, West Asia, and North Africa (Erskine et al. 2009). The seeds are a cheap source of protein with a high vitamin and mineral content, thus contributing to alleviation of hidden hunger. However, lentil yields are generally low. Lentil is commonly grown in developing countries on marginal lands with poor soil in rainfed environments (Gahoonia et al. 2006). Like many other crops that grow in Mediterranean and semiarid climates, lentil often faces terminal drought stress during the reproductive phase as a consequence of diminishing rainfall or plant available water and rising temperatures. In such cases, what is usually called "drought tolerance" could be, in part, the consequence of plant constitutive traits that affect how soil water is used when water is non-limiting to plant transpiration (Vadez et al. 2014). Until recently, the two water-conservation traits considered in this book have been missing in lentil drought studies, so this chapter mainly reports research that is currently underway.

In lentil, experiments conducted in Western Australia, with a Mediterranean-type climate, showed that seed yield was not correlated with total water use or with water use before flowering (Leport et al. 1998), but was positively correlated with post-flowering water use (Siddique et al. 2001). Under rainfed environments in

M.E. Ghanem (✉) • F. ez-zahra Kibbou
International Center for Agricultural Research in the Dry Areas (ICARDA), Crop Physiology Laboratory, North-Africa Platform, Rabat, Morocco
e-mail: M.Ghanem@cgiar.org

H. Marrou
UMR-SYTEM, Montpellier SupAgro, Montpellier, France

J. Guiguitant
International Center for Agricultural Research in the Dry Areas (ICARDA), Crop Physiology Laboratory, North-Africa Platform, Rabat, Morocco

UMR-SYTEM, Montpellier SupAgro, Montpellier, France

T.R. Sinclair (ed.), *Water-Conservation Traits to Increase Crop Yields in Water-deficit Environments*, SpringerBriefs in Environmental Science,
DOI 10.1007/978-3-319-56321-3_7

Nepal and Jordan, early-sown lentils produced greater yield as a result of adequate vegetative mass, longer grain-filling periods, and higher pod number (Neupane and Shrestha 1991; Shrestha 1996; Rahman et al. 2002). It is, however, surprising that these findings on seasonal dynamics of water use has not been investigated any further in lentil as to whether there is early-season soil water conservation so that more water is available to complete seed filling under drought conditions, occurring late in the growing season (Ludlow and Muchow 1990; Sinclair et al. 2005).

7.1 Water Conservation by Sensitivity to Soil Drying

Only now are there studies underway to investigate the transpiration response to soil drying in lentil (Kibbou et al. 2017, unpublished). The initial study was done using 17 diverse genotypes grown in India. Dry-down experiments were done in pots with mineral soil. The results for all genotypes were well described by the two-linear-segment model as illustrated by two divergent genotypes in Fig. 7.1. The R^2 for all genotypes was greater than 0.78 and only five genotypes were R^2 less than 0.85 (Table 7.1).

There was a very wide range in the FTSW threshold at which the decline in transpiration was initiated with values from 0.29 to 0.70 (Table 7.1). Eleven of the 17 genotypes had thresholds at 0.45 or greater, which are values greater than often

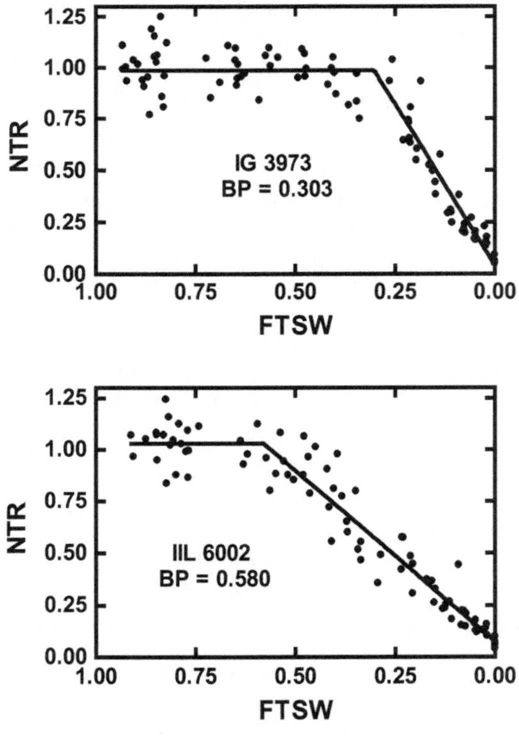

Fig. 7.1 Two-segment linear regressions of normalized transpiration rate (*NTR*) versus fraction of transpirable soil water (*FTSW*) for lentil genotypes IG 3793 and ILL 6002. The FTSW value at the breakpoint (*BP*) are shown for each genotype

Table 7.1 Lentil fraction transpirable soil water (FTSW) thresholds measured in the dry-down experiments of 17 diverse Indian genotypes

Genotype	Threshold	R^2
IG4258	0.289 a	0.784
IG 3973	0.303 a	0.925
VKS 16/11	0.369 a	0.857
IG 3964	0.373 abc	0.819
IG 2507	0.378 bc	0.929
FLIP2009-55 L	0.427 bcd	0.861
EC208362	0.450 bcd	0.825
P2016	0.450 cd	0.925
ILL 9997	0.457 cd	0.944
DPL-15	0.466 cd	0.900
ILL7663	0.480 d	0.953
JL 1	0.487 de	0.956
IPL98/193	0.488 de	0.934
DPL 53	0.500 de	0.854
ILL6002	0.580 def	0.936
ILL 9977	0.586 ef	0.814
IPL406	0.700 f	0.810

Thresholds followed by same letter are not significantly different based on 95% confidence interval

found in commercial genotypes of other species. These results indicate that there has been cryptic selection of lentil for the water-conservation trait with soil drying likely already occurring in genotypes that do well in more arid regions.

7.2 Water Conservation by Sensitivity to Vapor Pressure Deficit

No information is currently available on the transpiration response of lentil genotypes to vapor pressure deficit (VPD) although we have now completed such observations (Guiguitant et al. 2017, submitted). The purpose of this study was to determine if there are lentil genotypes that exhibit a breakpoint in their response to increasing VPD above which there is little or no further increase in transpiration rate. The 17 genotypes examined for their response to soil drying were also tested for their response to VPD. Transpiration rates of these selected lines were initially tested under natural daily VPD variation in Hyderabad, India. Surprisingly, none of the lines showed a clear breakpoint in their TR response to VPD (Guiguitant et al. 2017, submitted). However, the natural daily cycle of VPD did not expose the plants to VPD greater than 3 kPa.

To expand the genetic range of tested lentil genotypes, a selection of genotypes from a pool of 200 landraces was undertaken (Kibbou et al. 2017, unpublished). The

200 genotypes were identified in ICARDA's focused identification of germplasm strategy (FIGS). All 200 landraces were field prescreened for yield in two consecutive years in Marrakech, Morocco. A late sowing was used to ensure the plants would be subjected to drought stress. From the initial field screens, ten landraces originating from Pakistan, Nepal, and Ethiopia were selected for evaluation of transpiration response to VPD. Fourteen lentil genotypes coming from the USA and Canada were also evaluated in the same experiment.

Those genotypes selected in the prescreen were tested over a range of VPD from 1.0 to 3.8 kPa in semi-controlled conditions (greenhouse) by artificially drying the air at a stable temperature (24 °C). Similar to the results previously obtained by Guiguitant et al. (2017), none of the examined genotypes showed a clear breakpoint in their TR response to VPD (Kibbou et al. 2017, unpublished).

Since the range of VPD to which the plants were subjected in the second set of experiments was also somewhat limited (VPD < 3.8 kPa), a third set of experiments in a controlled environment were undertaken. In third set of experiments, the VPD range for which transpiration response was measured was for 2 to 6 kPa by maintaining a relatively high background temperature of >35 °C (Guiguitant et al. 2017, submitted). Measurements under this extended range of VPD showed a breakpoint in increasing transpiration rate at about 3.4 kPa in all lines except one (Guiguitant et al. 2017, submitted) (Fig. 7.2). At this point, it is not known if the high temperature used in this third set of experiments influenced the transpiration response at higher VPD values.

Altogether, these results indicated that lentil appears to have the genetic capacity to regulate their transpiration to VPD but that the limited-transpiration trait exists in lentil genotypes tests at very high VPD. Therefore, for most field conditions, the trait for response to VPD would not be very useful in terms of impact on water conservation. These preliminary genetic screenings support the need for more experimental screenings to evaluate genetic material of lentil in order to find genotypes that express a breakpoint at much lower VPD.

7.3 Extent of Yield Advantage from the Water-Saving Traits

The impact of the water-saving traits on lentil productivity is likely to vary across growing seasons with geography, environment type, and level of expression of the trait. If there is late-season rainfall or if the decrease of photosynthetic rate due to partial stomatal closure is too high, the benefit of conserved water would not be obtained; thus, lines with water-conservation traits lead to equal or lower yield than lines that do not express these traits (Sinclair et al. 2016). The geographical extent of the advantage and probability for lentil yield increase cannot yet be resolved due to limited experimental data.

However, simulation studies using the lentil version of the simple simulation model (SSM, Soltani and Sinclair 2012) do offer insights. The model was used to

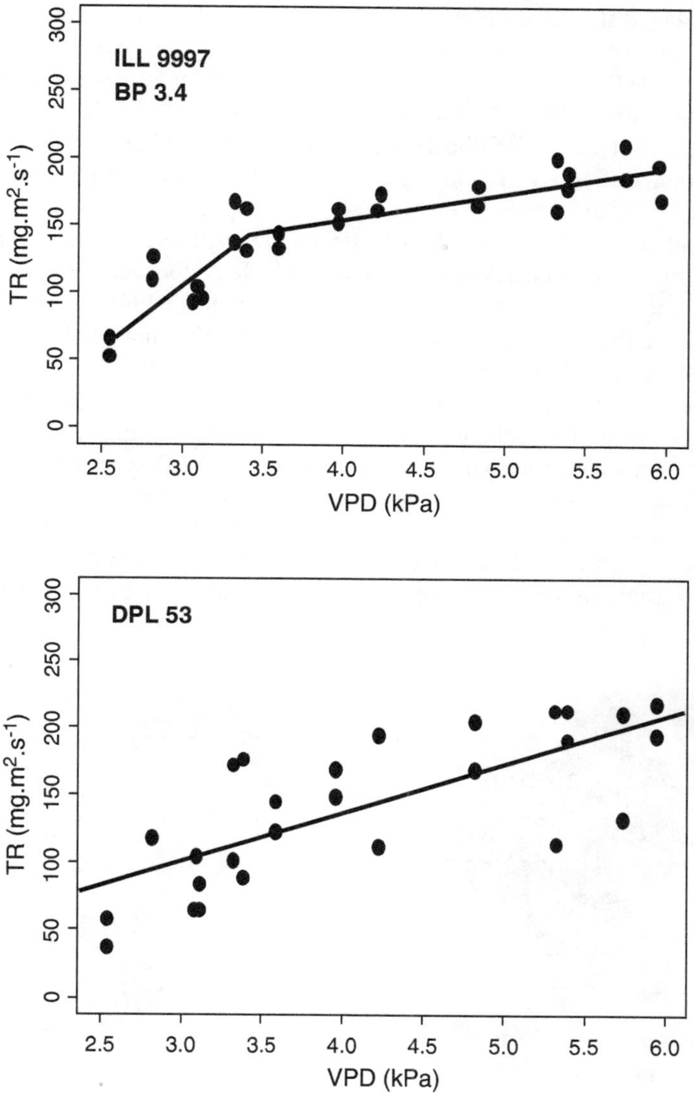

Fig. 7.2 Examples of two-segment lentil genotypes (ILL997 and DPL53) with differing responses to increasing vapor pressure deficit (*VPD*) when tested in a growth chambers over a wide range VPD (Guiguitant et al. 2017, submitted)

evaluate variation in phenological development and yield of lentil in a range of environments, with different rainfall patterns, in the Middle East (Ghanem et al. 2015a), investigating the roles of changing phenology and sowing dates on the possible expansion of lentil culture in East Africa (Ghanem et al. 2015b). More recently, the SSM model was used to simulate lentil yield in South Asia in response to the introduction of the VPD water-conservation trait. Daily weather data were generated for 447 different locations spanning over India, Afghanistan, Bangladesh, Pakistan, and Nepal for 30 years for each location using the Weather Research and Forecasting (WRF) tool (Ghanem et al. 2015b). Based on the experimental results, the standard simulations were done by including a VPD breakpoint at 3.4 kPa (Guiguitant et al. 2017, submitted). Simulations were also done for the case where transpiration was insensitive to high VPD (no VPD threshold) and a low VPD threshold (1.1 kPa).

The yield results for the two hypothetical VPD responses were compared to the 3.4 kPa simulation results. A probability was calculated for each simulated location of a yield increase as a result of changing the VPD threshold based on 30 years of simulation. In the results for simulations with no VPD threshold, only in a few locations was the probability of yield increase greater than 50%, i.e., this alteration resulted in yield loss in the majority of cases (Fig. 7.3a), when the threshold was lost. In most cases, the probability of yield increase was less than 25%.

In contrast, simulations with the breakpoint for the VPD at 1.1 kPa as compared to 3.4 kPa showed wide areas for high probability of yield increase, particularly in Central India and in Bangladesh (Fig. 7.3b). The largest average yield increase was 250 g m^{-2} located in the more drought-prone environments (data not shown). Outcomes from this simulation study provide evidence for the plausible role of the

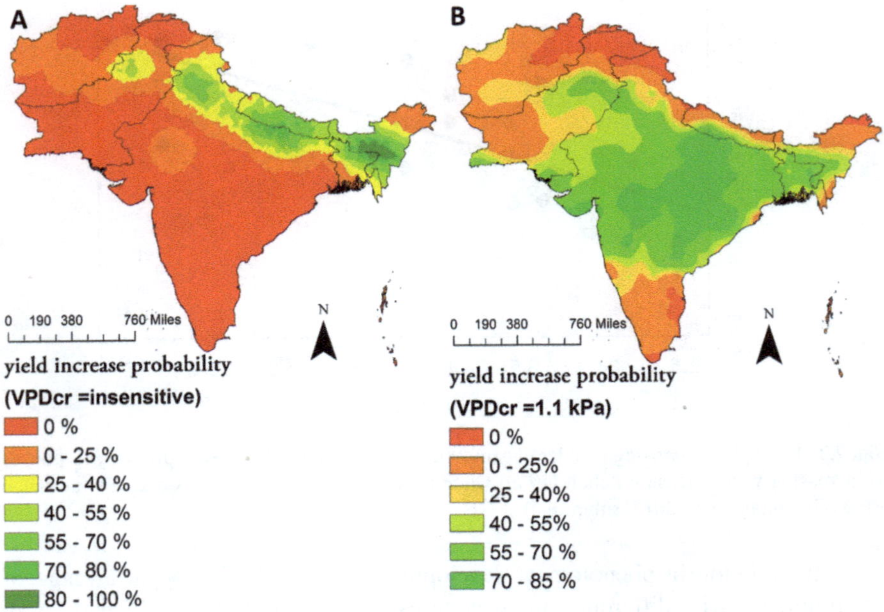

Fig. 7.3 Results of simulations of lentil across South Asia with inclusion of the limited-transpiration trait (Guiguitant et al. 2017, submitted). Probability of grain yield increases as compared to a lentil genotype that shows a VPD breakpoint at 3.4 kPa for hypothetical cultivars that (**a**) express a linearly increasing transpiration rate with increasing VPD (VPD insensitive) and (**b**) express a VPD breakpoint set at 1.1 kPa (Guiguitant et al. 2017, submitted)

limited-transpiration trait under high VPD in future lentil genetic improvement and confirmed that a search for germplasm with a low breakpoint needs to be made (Guiguitant et al. 2017, submitted).

References

Erskine W, Muehlbauer FJ, Sarker A et al (2009) Global production supply and demand. In: CABI (ed) The lentil: botany, production and uses. CABI, Wallingford, Oxfordshire

Gahoonia TR, Ali O, Sarker A et al (2006) Genetic variation in root traits and nutrient acquisition of lentil genotypes. J Plant Nutr 29:643–655

Ghanem ME, Marrou H, Soltani A et al (2015a) Lentil variation in phenology and yield evaluated with a model. Agron J 107:1967–1197

Ghanem ME, Marrou H, Biradar C et al (2015b) Production potential of lentil (*Lens culinaris* Medik.) in East Africa. Agric Syst 137:24–38

Guiguitant J, Marrou H, Vadez V et al (2017) Relevance of limited-transpiration trait for lentil (*Lens culinaris Medik.*) in South Asia. Field Crop Res (accepted in press)

Kibbou F, Marrou H, Ghanem ME et al (2017) Genetic variability of transpiration response to soil drying among lentil (*Lens culinaris* Medik.) cultivars. Crop Sci (submitted)

Leport L, Turner NC, French RJ et al (1998) Water relations, gas exchange and growth of cool-season grain legumes in a Mediterranean-type environment. Eur J Agron 9:295–303

Ludlow MM, Muchow RC (1990) A critical evolution of traits for improving crop yields in water-limited environments. Adv Agron 43:107–153

Neupane RK, Shrestha R (1991) Report on varietal and agronomical studies on lentil. Paper presented at the Winter Crop Workshop in Nepal Agricultural Research Council (NARC), Kathmandu, Nepal, 16–20 Sept

Rahman A, Tawaha M, Turk MA (2002) Effect of dates and rates of sowing on yield and yield components of lentil (*Lens culinaris* Medik.) under semi-arid conditions. Pak J Biol Sci 5:531–532

Shrestha R (1996) Effect of planting dates and seed rates on lentil (var. Simal) at Khumaltar (1993/94–1995/96). In: National Winter Crop Workshop. Regional Agricultural Research Station, Bhairahawa. Nepal Agricultural Research Council, Bhairahawa, Lumbini, Nepal

Siddique KHM, Regan KL, Tennant D et al (2001) Water use and water use efficiency of cool season grain legumes in low rainfall Mediterranean type environments. Eur J Agron 15:267–280

Sinclair TR, Hammer GL, van Oosterom EJ (2005) Potential yield and water-use efficiency benefits in sorghum from limited maximum transpiration rate. Funct Plant Biol 32:945–952

Sinclair TR, Devi JM, Carter TE (2016) Limited-transpiration trait for increased yield for water-limited soybean:from model to phenotype to genotype to cultivars. In: Yin X, Struik PC (eds) Crop systems biology. Springer, Cham, pp 129–143

Soltani A, Sinclair TR (2012) In: CABI (ed) Modeling physiology of crop development, growth and yield. CABI, Wallingford, Oxfordshire

Vadez V, Kholova J, Medina S et al (2014) Transpiration efficiency: new insights into an old story. J Exp Bot 64:6141–6153

Chapter 8
Maize

Avat Shekoofa and Sunita Choudhary

Drought is often the most important stress behind yield losses and impacts nearly all crop production for at least a limited period of time across the US corn belt (Cooper et al. 2014). Maize (*Zea mays* L.) yield is often restricted by low soil water availability, particularly late in the growing season. Total annual crop damage from drought in the United States has been estimated at \$6 to \$8 billion (Federal Emergency Management Agency 1995). In the 3-year period from 2008 to 2010, the United States produced 39% of global maize. The maize yield reduction in 2012 due to drought stress relative to previous five non-drought years was 21% and the price change was 53% (Boyer et al. 2013). An important option for enhancing crop yield on limited water is to develop cultivars that are conservative in their use of soil water (Sinclair et al. 2010; Shekoofa et al. 2016).

8.1 Decrease in Transpiration with Progressive Soil Drying

An early simulation study with maize (Sinclair and Muchow 2001) showed that a threshold for the decrease in transpiration rate at high soil water content can have a beneficial impact on yield under drought conditions. The simulations were for Columbia, MO, over 20 growing seasons. By increasing the fraction of transpirable soil water (FTSW) threshold for the decrease in transpiration rate from 0.29 to 0.44,

---

A. Shekoofa (✉)
Department of Plant Sciences, University of Tennessee Knoxville,
605 Airways Blvd, Jackson, TN 38301, USA
e-mail: ashekoof@utk.edu

S. Choudhary
Multi-Crop Research Center, Pioneer Hi-Bred Private Limited, Hyderabad, India
e-mail: sunita963@gmail.com

T.R. Sinclair (ed.), *Water-Conservation Traits to Increase Crop Yields in Water-deficit Environments*, SpringerBriefs in Environmental Science,
DOI 10.1007/978-3-319-56321-3_8

Fig. 8.1 Daily normalized transpiration ratio (*NTR*) response to the fraction of transpirable soil water (*FTSW*) of individual plants tested for two maize hybrids [(**a**) Hybrid#2 BP=0.37 (**b**) Hybrid#17 BP=0.59] that showed contrasting FTSW thresholds. The breakpoints (*BP*) at which NTR began to decrease are given in each panel (Gholipoor et al. 2012)

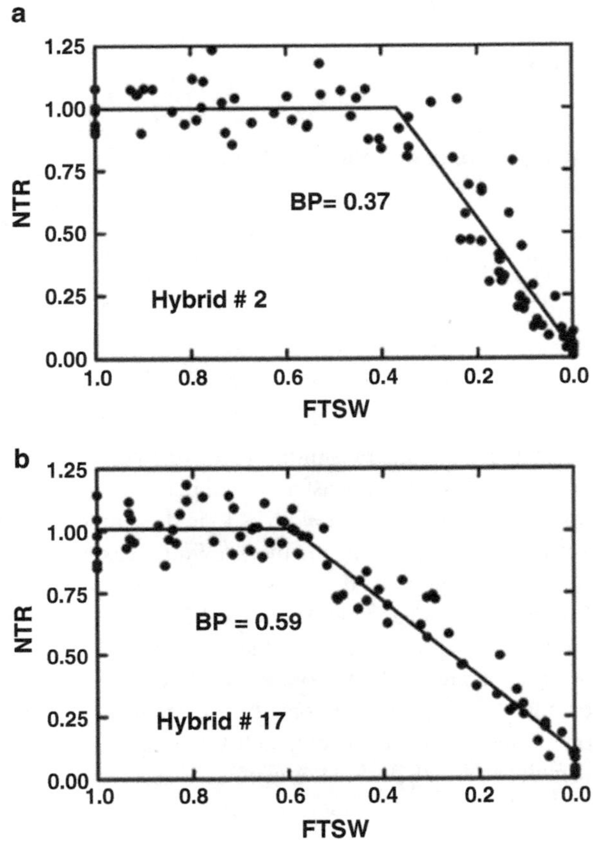

large yield increases were simulated in three of 20 growing seasons. There was little negative impact of the trait on yield in the remaining seasons.

Gholipoor et al. (2012) showed that FTSW thresholds in the soil drying cycle varied among maize hybrids from 0.60 to 0.33. However, these thresholds were not uniformly associated with the level of drought resistance in the field. Nevertheless, this study demonstrated a high FTSW threshold corresponded with drought resistance observed in some modern maize germplasm (hybrids #7, 17, 24, 27, and 32 in Fig. 8.1). Overall, these results demonstrated important genetic variation among hybrids in the relative value of the FTSW threshold during soil drying and offered a potential for yield increase in environments where soil water deficits are common.

8.2 Transpiration Sensitivity to Vapor Pressure Deficit

Another approach to achieve water conservation is to have limited-transpiration rate (TR_{lim}) under high atmospheric VPD as a result of partial closure of stomata. Partial stomatal closure would occur when water transfer to guard cells cannot match the transpiration rate (TR) that results from fully open stomata (Brodribb and Jordan 2008; Yang et al. 2012; Shekoofa et al. 2013, 2014; Choudhary et al. 2013). Therefore, under high VPD conditions partial stomatal closure, which can develop during the midday, water would be conserved for use later in the growing season.

Gholipoor et al. (2013) found that like other species, there are individual maize hybrids that exhibited a breakpoint in the increase in TR with increasing VPD. The hybrids were studied in two sets of experiments. Twenty hybrids were studied in their first set of experiments, and 13 additional hybrids were studied in a second set of experiments along with eight hybrids selected for retest. In addition Yang et al. (2012) reported results for additional two hybrids, so results have been reported for a total of 35 hybrids. Among 35 temperate maize hybrids, a total of 11 hybrids were identified expressing the breakpoint in transpiration rate with increasing VPD.

Results for eight hybrids that were included in both sets of experiments showed consistency in the expression of a breakpoint in the transpiration rate at high VPD. All six hybrids identified with the transpiration breakpoint in the first set of experiments were confirmed to express the limited transpiration in the second set of experiments. Similarly, the two hybrids with a linear response over all VPD expressed this response in the second set of experiments (Gholipoor et al. 2013). These results indicated that the overall TR_{lim} response of individual hybrids for the expression of the breakpoint was repeatable across experiments (Gholipoor et al. 2013).

Yang et al. (2012) confirmed the TR_{lim} trait in four of the maize hybrids identified by Gholipoor et al. (2013) as expressing the trait. However, in the tests of Yang et al., measurements were made at temperatures lower than 32 °C, which was the temperature used by Gholipoor et al. The breakpoint for TR_{lim} shifted to lower values for both VPD and transpiration rate for measurements at 25 °C as compared to 30 °C (Table 8.1). The shift in the breakpoint is likely caused by temperature-induced factors affecting water supply rate to the transpiration sites in the leaf surface (Yang et al. 2012).

8.2.1 Influence of High Temperature

A major issue for maize is expression of the TR_{lim} trait at high temperatures (>32 °C) that are likely to be relevant to the environmental conditions in which maize is commonly grown. To test the possible influence of temperature on expression of TR_{lim}, two pairs of tropical maize hybrids contrasting for drought tolerance and expression of TR_{lim} were grown in a matrix of temperature and VPD treatments (Choudhary et al., unpub. data). The four treatment combinations were two temperatures (29 °C and 35 °C) and two VPDs (1.7 kPa and 3.5 kPa). To evaluate for expression of TR_{lim}, TR were measured on a single day over VPD ranging from 1.1 to 3.2 kPa at the respective growth temperatures. The VPD breakpoint of the two hybrids that expressed the breakpoint at 32 °C was also expressed in plants grown at 35 °C although the VPD of the breakpoint was higher for plants grown at 35 °C

Table 8.1 Vapor pressure deficit at breakpoint (BP) in transpiration rate with increasing vapor pressure over the range of measured temperatures for each maize hybrid (Shekoofa et al. 2016)

Hybrids	25	27.5	30	32	34	36	38
	BP (kPa)[a]	BP ± SE(kPa)	BP (kPa)[a]	BP (kPa)[b]	BP ± SE(kPa)	BP ± SE(kPa)	BP ± SE(kPa)
PHYBTS013	1.81	2.04 ± 0.36	2.13	1.74	2.26 ± 0.20	2.01 ± 0.42	Linear
PHYBTS007	–	–	–	1.86	2.45 ± 0.40	2.15 ± 0.40	Linear
PHYBTS010	1.75	1.62 ± 0.67	2.08	1.91	1.90 ± 0.50	1.18 ± 0.34	Linear
PHYBTS017	1.86	1.80 ± 0.44	2.19	2.01	1.95 ± 0.28	2.14 ± 0.47	2.60 ± 0.84
PHYBTS030	–	–	–	2.01	–	–	3.20 ± 0.24
PHYBTS004	–	2.29 ± 0.15	–	2.06	2.42 ± 0.10	1.62 ± 0.71	3.70 ± 0.53
PHYBTS027	–	–	–	2.11	–	–	2.26 ± 0.44
PHYBTS031	–	–	–	2.23	2.57 ± 0.25	2.62 ± 1.13	Linear
PHYBTS024	–	1.82 ± 1.00	–	2.40	2.40 ± 0.24	2.16 ± 0.30	Linear
PHYBTS021	1.72	1.87 ± 0.63	2.04	2.50	2.10 ± 0.26	1.23 ± 0.31	2.04 ± 1.31
PHYBTS029	–	–	–	2.50	–	–	2.05 ± 0.54
PHYBTS019	–	–	–	2.52	–	–	1.98 ± 0.28

[a]Yang et al. (2012)
[b]Gholipoor et al. (2013)

in comparison with 29 °C. For all four hybrids, TR was increased under the higher temperature irrespective of VPD.

Shekoofa et al. (2016) specifically examined the persistence of the TR_{lim} trait when maize plants were exposed to high temperatures. They examined the response of the hybrids originally identified by Gholipoor et al. (2013) over the temperature range from 27.5 to 38 °C. Seven of the hybrids continued to express the TR_{lim} trait over the whole temperature range (Table 8.1). These hybrids would, of course, be useful in those environments where expression of the TR_{lim} trait is consistently desired. On the other hand, five of the hybrids lost the trait between 36 °C and 38 °C (Table 8.1). Hybrids that lose the TR_{lim} trait might be appropriate in climates in which very high temperatures are frequent, potentially causing temperature stress; it may be advantageous to consider hybrids that lose the TR_{lim} trait at 38 °C rather than 32 °C or other temperatures below 38 °C (Table 8.1). Loss of TR_{lim} trait at very high temperature will result in greater TR and evaporative cooling under these potentially damaging conditions.

The sensitivity of TR_{lim} to temperature might be a result of changes in plant hydraulic conductivity. It has been shown that there can be temperature-dependent changes in both membrane permeability and water viscosity (Matzner and Comstock

2001; Sack and Holbrook 2006). In fact, Choudhary et al. (2014) found that the VPD breakpoint at 32 °C of 12 maize hybrids was positively correlated with indices of hydraulic conductance in both the leaves and roots. The loss of the VPD breakpoint at 38 °C in five hybrids may indicate an abrupt increase in hydraulic conductance at high temperatures in these hybrids.

8.2.2 Hydraulic Conductivity and Aquaporin Inhibitors

Root–shoot water balance is essential for plant survival and growth. Therefore, hydraulic conductivity through roots and leaves in particular could be critical in determining water flow through the plant. Previous studies have shown that drought-tolerant maize lines have better strategies for regulating hydraulics than drought-sensitive lines (Cochard 2002; Zhao et al. 2014).

The genetic variation in the TR_{lim} trait discussed previously was hypothesized to be due to differences in hydraulic conductance in the roots or leaves, or both (Gholipoor et al. 2013). Direct comparison of relative hydraulic conductivity in the roots and leaves across these maize hybrids was studied by Choudhary et al. (2014). They found that value of the VPD for the TR_{lim} breakpoint was correlated with each of the two indices of hydraulic conductance. That is, low VPD breakpoint was associated with low hydraulic conductivity both in leaves and roots. The commonality in leaves and roots indicated there may be a common, underlying, limiting mechanism in these two tissues. It was hypothesized that expression of similar aquaporin populations influencing hydraulic flow across membranes in the roots and leaves may account for the consistency in results across the indices of hydraulic conductivity.

In an effort to determine the possible involvement of aquaporins, Choudhary et al. (2015) examine TR response of 21 maize hybrids following treatment with inhibitor aquaporins. They treated both individual leaves and intact plants with cycloheximide (CHX) and four aquaporin inhibitors: silver ($AgNO_3$), gold ($HAuCl_4$), zinc ($ZnCl_2$), and mercury ($HgCl_2$). There was no discrimination among hybrids based on treatment with mercury or CHX. Segregation between hybrids for response to increasing VPD corresponded with differences in leaf response to silver and gold treatment and intact plant response to zinc. The highest correlation (R^2 = 0.90) was found between VPD breakpoint and TR response to silver treatment of leaves (Fig. 8.2). In their recent study, Shekoofa and Sinclair et al. (unpub. data) found that there also was a correlation (R^2 = 0.63) between decrease in transpiration rate (DTR%) of treated maize leaves with $AgNO_3$ and TR_{lim} trait (VPD breakpoint) even under 38 °C and high VPD. The results of these studies have led to the hypothesis that the insensitivity to silver and low conductivity indicated the absence of silver-sensitive aquaporins that allowed greater conductivity. These results indicate that $AgNO_3$ may be an effective initial screen for expression of the TR_{lim} trait under 32 °C and high VPD.

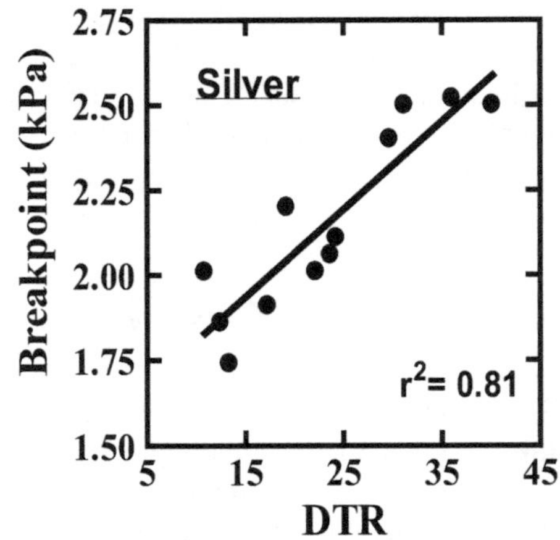

Fig. 8.2 Correlation between the response of maize leaves fed with silver nitrate (Ag) under 32°C and high VPD with TR_{lim} (VPD breakpoint) (Choudhary et al. 2014)

8.2.3 Extent of Yield Advantage

Stress conditions are associated with soil water deficit (Boyer 1982) and high water demand (Lobell et al. 2013, 2014) and are often manifested in decreased maize yields (Schlenker and Roberts 2009). Genotypic diversity among maize hybrids in the transpiration response to vapor pressure deficit (VPD) indicates that a TR_{lim} trait may contribute to improved drought tolerance and yield in maize. The complexity associated with genotype × environment × management interactions can be explored in a quantitative assessment using a simulation model.

A simulation study was conducted to assess the likely effect of genotypic variation in TR_{lim} rate on yield performance of maize at a regional scale in the United States. Messina et al. (2015) demonstrated that the TR_{lim} trait (vapor pressure deficit breakpoint of 2 kPa) can result in improved maize performance in drought-prone environments and that the impact of the trait on maize productivity varies with geography, environment type, expression of the trait, and plant density. The largest average yield increase was simulated for drought-prone environments (135 g m^{-2}), while a small yield penalty was simulated for environments where water was not limiting (−33 g m^{-2}) (Fig. 8.3).

The simulation results are consistent with the experience of DuPont Pioneer in marketing their AQUAmax maize hybrids. These hybrids express the TR_{lim} trait (Gaffney et al. 2015) and were shown to have greater soil water contents than normal hybrids through the growing season (Cooper et al. 2014). Gaffney et al. (2015) compared the yield of AQUAmax hybrids with non-AQUAmax hybrids in small plot experiments and in hundreds of on-farm strip trials in water-limited environments. The average yield advantage of the AQUAmax hybrids over 3 years at 53 water-limited locations was 4.9%. In the on-farm trials involving hundreds of locations, the average yield advantage in 3 years was 7.3, 9.7, and 3.3% in 2011, 2012, and 2013, respectively (Gaffney et al. 2015).

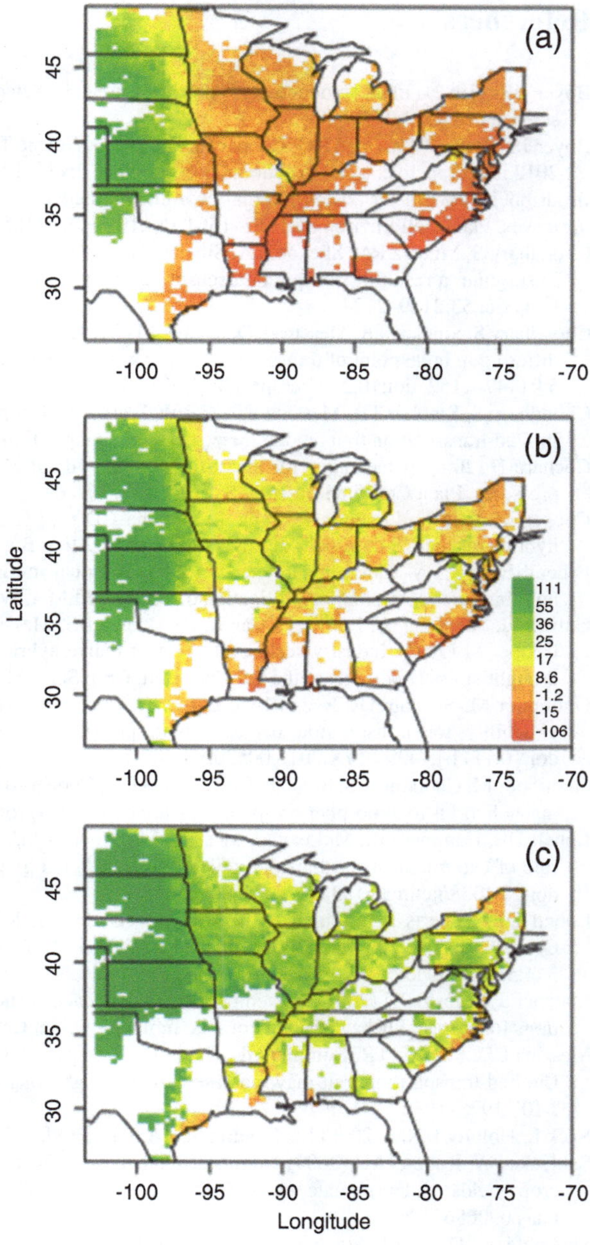

Fig. 8.3 Spatial distributions of the simulated yield difference (g m^{-2}) associated with a change in transpiration response to limited (vapor pressure deficit breakpoint of 2 kPa) from the nonlimited reference with an exceedance probability of (**a**) 25%, (**b**) 50%, and (**c**) 75% (Messina et al. 2015)

References

Boyer JS (1982) Plant productivity and environment. Science 218:443–448. doi:10.1126/science.218.4571.443

Boyer JS, Byrne P, Cassman KG, Cooper M, Delmer D, Greene T et al (2013) The U.S. drought of 2012 in perspective: a call to action. Glob Food Sec 2:139–143. doi:10.1016/j.gfs.2013.08.002

Brodribb TJ, Jordan GJ (2008) Internal coordination between hydraulic and stomatal control in leaves. Plant Cell Environ 31:1557–1564. doi:10.1111/j.1365-3040.2008.01865.x

Choudhary S, Mutava RN, Shekoofa A, Sinclair TR, Vara Prasad PV (2013) Is the stay-green trait in sorghum a result of transpiration sensitivity to either soil drying or vapor pressure deficit? Crop Sci 53:2129–2134

Choudhary S, Sinclair TR, Messina CD, Cooper M (2014) Hydraulic conductance in maize hybrids differing in breakpoint of transpiration response to increasing vapor pressure deficit. Crop Sci 54:1147–1152. doi:10.2135/cropsci2013.05.0303

Choudhary S, Sinclair TR, Messina CD, Cai W, Warner D, Cooper M (2015) Inhibitor screen for limited-transpiration trait among maize hybrids. Environ Exp Bot 109:161–167

Cochard H (2002) A technique for measuring xylem hydraulic conductance under high negative pressures. Plant Cell Environ 25:815–819

Cooper M, Gho C, Leafgren R, Tang T, Messina CD (2014) Breeding drought-tolerant maize hybrids for the US corn-belt: discovery to product. J Exp Bot 65:6191–6204

Federal Emergency Management Agency (1995) National mitigation strategy—partnerships for building safer communities. Mitigation directorate\FEMA, Washington, DC

Gaffney J, Schussler W, Loffler C, Cai W, Paszkiewicz S, Messina CD, Groeteke J, Keaschall J, Cooper M (2015) Industry-scale evaluation of maize hybrids selected for increased yield in drought-stress conditions of the US Corn Belt. Crop Sci 55:1608–1618

Gholipoor M, Sinclair TR, Raza MAS, Loffler C, Cooper M, Messina CD (2012) Maize hybrid variability for transpiration decrease with progressive soil drying. J Agron Crop Sci. doi:10.1111/j.1439-037X.2012.00530.x

Gholipoor M, Choudhary S, Sinclair TR, Messina CD, Cooper M (2013) Transpiration response of maize hybrids to atmospheric vapour pressure deficit. J Agron Crop Sci 199:155–160

Lobell DB, Hammer GL, McLean G, Messina CD, Roberts MJ, Schlenker W (2013) The critical role of extreme heat for maize production in the United States. Nat Clim Chang 3:497–501. doi:10.1038/nclimate1832

Lobell DB, Roberts MJ, Schlenker W, Braun N, Little BB, Rejesus RM, Hammer GL (2014) Greater sensitivity to drought accompanies maize yield increase in the US Midwest. Science 344:516–519. doi:10.1126/science.1251423

Matzner S, Comstock J (2001) The temperature dependence of shoot hydraulic resistance: implications for stomatal behavior and hydraulic limitation. Plant Cell Environ 24:1299–1307

Messina CD, Sinclair TR, Hammer GL, Curan D, Thompson J, Oler Z, Gho C, Cooper M (2015) Limited-transpiration trait may increase maize drought tolerance in the US Corn Belt. Agron J 107:1978–1986

Sack L, Holbrook NM (2006) Leaf hydraulics. Annu Rev Plant Biol 57:361–381

Schlenker W, Roberts MJ (2009) Nonlinear temperature effects indicate severe damages to U.S. crop yields under climate change. Proc Natl Acad Sci 106:15594–15598. doi:10.1073/pnas.0906865106

Shekoofa A, Devi JM, Sinclair TR, Holbrook CC, Isleib TG (2013) Divergence in drought-resistance traits among parents of recombinant peanut inbred lines. Crop Sci 53:2569–2576. doi:10.2135/cropsci2013.03.0153

Shekoofa A, Balota M, Sinclair TR (2014) Limited-transpiration trait evaluated in growth chamber and field for sorghum genotypes. Environ Exp Bot 99:175–179. doi:10.1016/j.envexpbot.2013.11.018

Shekoofa A, Sinclair TR, Messina CD, Cooper M (2016) Variation among maize hybrids in response to high vapor pressure deficit at high temperatures. Crop Sci 56:392–396. doi:10.2135/cropsci2015.02.0134
Sinclair TR, Muchow RC (2001) System analysis of plant traits to increase grain yield on limited water supplies. Agron J 93:263–270
Sinclair TR, Messina CD, Beatty A, Samples M (2010) Assessment across the United States of the benefits of altered soybean drought traits. Agron J 102:475–482. doi:10.2134/agronj2009.0195
Yang Z, Sinclair TR, Zhu M, Messina CD, Cooper M, Hammer GL (2012) Temperature effect on transpiration response of maize plants to vapour pressure deficit. Environ Exp Bot 78:157–162. doi:10.1016/j.envexpbot.2011.12.034
Zhao W, Sun Y, Kjelgren R, Liu X (2014) Response of stomatal density and bound gas exchange in leaves of maize to soil water deficit. Acta Physiol Plant 37:1–9

Chapter 9
Sorghum

Sunita Choudhary and Jana Kholová

Grain sorghum is grown on about 45 M ha worldwide with an average yield of 1.4 t ha^{-1} and ranks as the fifth most important cereal worldwide (http://faostat3.fao.org/download/Q/QC/E). Sorghum production systems encompass predominantly semiarid tropical regions of South Asia and Africa. Sorghum grain is an important staple food of the poverty-threatened farming populations of these regions. Beyond human consumption, sorghum grain and stover are also used as animal feed, source of bioenergy, lignin, and sweet stem varieties for ethanol production. Since sorghum is relatively more adapted to various abiotic constraints (heat, drought, salinity, low soil fertility) compared to other major cereals, most of its cultivated areas are pushed to the marginal land and given minimum inputs. Despite the general crop resilience of sorghum, water deficit accounts for significant losses in grain yield in many cases (e.g., Kholová et al. 2013). From this point of view, adapting the crop agronomy and developing the genotypes with better fitness to these water-limited production systems could increase crop production and improve the livelihood of resource-limited farming communities in developing countries.

S. Choudhary (✉)
Multi-Crop Research Center, Pioneer Hi-Bred Private Limited, Hyderabad, India
e-mail: sunita963@gmail.com

J. Kholová
International Crops Research Institute for the Semi-Arid Tropics (ICRISAT), Greater Hyderabad, Telangana, India
e-mail: j.kholova@cgiar.org

T.R. Sinclair (ed.), *Water-Conservation Traits to Increase Crop Yields in Water-deficit Environments*, SpringerBriefs in Environmental Science,
DOI 10.1007/978-3-319-56321-3_9

9.1 Crop Responsiveness to Soil Drying

9.1.1 Sorghum Growth and Transpiration Response to Soil Drying

Regulation of stomatal conductance and leaf expansion are the early responses of plants to progressing soil water deficits (Jones and Corlett 1992; Turner 1997). Tight regulation of canopy growth and conductance upon soil drying allows plants to conserve the remaining soil moisture and extend their period of physiological activity during drought. In sorghum, considerable genetic variation was reported in the soil water threshold (fraction of transpirable soil water; FTSW) when plants began to limit transpiration and mass growth (Ghoolipoor et al. 2012; Choudhary et al. 2013a, b; Choudhary and Sinclair 2014a). Gholipoor et al. (2012) and Choudhary et al. (2013a) studied 16 and 10 genotypes (breeding lines and converted tropical lines from a sorghum panel), respectively, and found that the FTSW threshold for initiation of limit transpiration rate ranged from 0.32 to 0.48 (Table 9.1).

9.1.2 Simulated Impact on Grain and Stover Yield

Kholová et al. (2014) undertook simulations with the APSIM model to evaluate the possible influence in sorghum of the sensitivity of decrease in transpiration rate and leaf area development. Simulations incorporating sensitivity of transpiration rate to soil drying was done by altering assumed soil-root hydraulic values. For the environmental scenarios simulated, stover yield was generally increased as a result of greater transpiration sensitivity to soil drying but slightly decreased as a result of increased sensitivity in leaf area expansion. However, there was little consistent impact on grain yield as a result of alteration in either transpiration rate or leaf area development in response to soil drying.

9.2 Plant Responsiveness to Atmospheric Drought (Vapor Pressure Deficit)

9.2.1 Limited Transpiration Rate Under High Vapor Pressure Deficit in Sorghum

One of the recently identified constitutive responses in sorghum to atmospheric vapor pressure deficit (VPD) is the plant ability to partially close their stomata above a threshold VPD. The partial closure of stomata results in limited-transpiration rate (amount of water transpired per unit of leaf area per unit of time) upon exposure

Table 9.1 Results of progressive soil drying (dry down) experiment for 16 sorghum genotypes (Gholipoor et al. 2012) and 10 sorghum breeding lines (Choudhary et al. 2013a): initial transpiration rate during first well water stage, days to reach endpcint of dry down, and FTSW threshold and confidence interval in the dry down

Genotype	Transpiration (g plant^{-1}d^{-1})	Days to endpoint of dry down	FTSW threshold for transpiration decline	Confidence interval ($p < 0.05$)
Gholipoor et al. (2012)				
SC982	42.1 ± 1.5	16.9 ± 0.2	0.319	0.28–0.35
BTX623	40.8 ± 1.6	16.5 ± 0.2	0.324	0.28–0.36
SC979	42.4 ± 1.7	16.0 ± 0.2	0.331	0.30–0.36
SC1074	39.3 ± 1.4	16.9 ± 0.2	0.336	0.30–0.37
SC599	44.5 ± 1.6	14.9 ± 0.2	0.358	0.32–0.40
SC630	45.8 ± 1.3	15.3 ± 0.2	0.360	0.33–0.39
SC1047	47.5 ± 1.3	14.4 ± 0.2	0.361	0.32–0.40
BTX378	51.2 ± 1.6	14.3 ± 0.2	0.364	0.32–0.40
SC803	47.2 ± 1.6	14.8 ± 0.2	0.373	0.34–0.41
TX7078	51.0 ± 1.7	13.5 ± 0.2	0.383	0.35–0.42
B35	49.3 ± 1.6	14.0 ± 0.2	0.384	0.35–0.42
TX3042	51.2 ± 1.6	13.7 ± 0.2	0.387	0.35–0.42
SC489	54.0 ± 1.8	13.0 ± 0.2	0.449	0.41–0.48
SC299	53.0 ± 2.0	13.2 ± 0.2	0.469	0.42–0.52
Hegari	55.8 ± 2.0	13.3 ± 0.2	0.473	0.43–0.52
DK28	54.0 ± 2.0	13.0 ± 0.2	0.480	0.44–0.52
Choudhary et al. (2013a)				
SC599	69.4 ± 6.5	10.0 ± 0.2	0.376 a	0.34–0.41
PI 889012	103.8 ± 5.1	9.0 ± 0.2	0.384 a	0.36–0.40
SC35	118.9 ± 6.9	9.0 ± 0.2	0.399ab	0.37–0.43
BQL41	56.6 ± 1.7	10.0 ± 0.2	0.402 ab	0.37–0.43
Tx436/07MN	66.7 ± 7.9	12.0 ± 0.2	0.452 b	0.42–0.49
R-45	56.9 ± 3.6	11.0 ± 0.2	0.476 b	0.43–0.52
Tx623	62.8 ± 4.4	11.0 ± 0.2	0.376 a	0.34–0.41
Tx436	53.3 ± 1.1	11.0 ± 0.2	0.408 ab	0.37–0.45
Tx436	68.8 ± 3.6	12.0 ± 0.2	0.416 ab	0.38–0.45
Tx7000	57.4 ± 2.4	10.0 ± 0.2	0.480 b	0.45–0.51

to high VPD. Sensitivity of stomatal conductance to increasing VPD was initially observed in sorghum (and potato) by Bunce (2003) under field conditions. The existence of genotypic variation for transpiration rate (TR) under high VPD conditions in sorghum was more extensively documented recently (Gholipoor et al. 2010; Choudhary et al. 2013a; Shekoofa et al. 2014). Gholipoor et al. (2010) showed that out of 26 sorghum genotypes from an association panel (Casa et al. 2008), 17 genotypes expressed limited TR between 1.62 and 2.72 kPa with very little or no increase in TR upon further VPD increment (Table 9.2). Shekoofa et al. (2014) further added that assessment of plant transpiration responsiveness obtained from controlled

Table 9.2 Transpiration rate vs. VPD regression results [breakpoint and slopes of two linear regression segments (mg H_2O m^{-2} s^{-1} kPa^{-1})] for 17 genotypes out of 26 that were found to fit the two-segment regression model (Gholipoor et al. 2010)

Genotype	Breakpoint (kPa)	Slope 1	Slope 2
SC982	1.62 ± 0.23	50 ± 12.6	6.3 ± 4.8
SC1074	1.99 ± 0.17	45.8 ± 9.5	1 ± 4.1
BTX623	2.05 ± 0.18	56.3 ± 10.7	10.4 ± 5.2
SC979	2.06 ± 0.18	47.7 ± 7.8	0 ± 6.5
BTX2752	2.08 ± 0.20	52 ± 9	11 ± 6.7
SC1019	2.08 ± 0.25	43.8 ± 8.1	1.3 ± 10.1
SC630	2.16 ± 0.31	40.1 ± 9.7	−1.5 ± 8.2
SC599	2.23 ± 0.27	35.5 ± 7	3.4 ± 8.4
SC1047	2.24 ± 0.23	32.2 ± 5.6	1.1 ± 6.5
SC803	2.29 ± 0.19	39.8 ± 6.5	−6.1 ± 6.8
BTX3197	2.29 ± 0.29	38.8 ± 6.5	3.2 ± 9.2
SN149	2.30 ± 0.19	38.2 ± 5.2	−1.8 ± 8.8
Macia	2.51 ± 0.27	35.7 ± 3.8	4 ± 12.4
BTX378	2.55 ± 0.39	30.2 ± 3.3	11.3 ± 11
B35	2.61 ± 0.18	40.3 ± 6.8	2.4 ± 7.7
TX3042	2.69 ± 0.23	33.5 ± 4.6	−0.5 ± 11.4
TX7078	2.72 ± 0.23	37.7 ± 6	1.4 ± 10.2

conditions related to field observations. This means that data obtained in controlled environments could hypothetically be used to do genotype selections and extrapolated to field conditions.

Kassahun et al. (2010) found that variability in TR responsiveness to VPD was associated with particular sorghum stay-green QTL introgressions. Choudhary et al. (2013a) and Kholová et al. (2014) also studied genotypes diverging in their expression of the stay-green trait and found variation in their TR responsiveness to VPD. Thresholds for limiting transpiration rate were found between 1.17 and 2.91 kPa in these genotypes, and it was suggested that this trait might be associated with the stay-green phenotype in sorghum. It was also found that transpiration sensitivity to increasing VPD led to higher transpiration efficiency in water use, measured gravimetrically over the entire crop cycle (Aparna et al. 2017, under review).

The impact of variability in TR responsiveness has been studied for sorghum using simulation models. Sinclair et al. (2005) simulated the potential benefit of the maximum-transpiration trait for sorghum over 100 years at four locations in Australia. The simulation results showed yield increases due to the trait at yield levels less than 4.5 t ha^{-1} and only small decreases in yield above this level. These simulation results are similar to the observed response reported by Jordan et al. (2012) for genotypes expressing the stay-green trait.

Kholová et al. (2014) simulated the effect of limited-transpiration trait on crop production across post-rainy sorghum production systems in India. The model predicted that variability for limited-transpiration rate available in Indian germplasm has high probability to consistently improve stover and grain yield across most of the target production regions. This is possible because the limited TR improves not only early-season water conservation but also transpiration efficiency. Nevertheless, one of the trade-offs possibly associated with lower TR is the restriction of mass accumulation during the high VPD portions of the day when stomata are partially closed. Before application of the limited-transpiration trait in sorghum, it is necessary to analyze carefully the location-specific associated trade-offs.

9.2.1.1 Environmental Conditions Affect Plant Responsiveness to VPD

Earlier, short-term studies have shown temperature affects on stomatal conductance and photosynthesis (e.g., Woledge et al. 1989). Since sorghum is frequently grown under high-temperature conditions, it is important to understand how the TR responsiveness changes in plants grown at higher temperatures. Riar et al. (2015) found that plants could lose their TR responsiveness to VPD or shift their range of TR responses to higher VPD thresholds (e.g., 2.5 to 3.0 kPa) if grown in high-temperature regimes (>37 °C).

Resolution of the high-temperature impact on the limited transpiration trait is important because it may have a direct effect on crop water use and, therefore, crop adaptation to water-limited environments (Parent et al. 2010). At the same time, loss of the transpiration sensitivity to VPD allows greater transpiration at high temperature and greater transpiration cooling which is an important mechanism of high-temperature stress adaptation. Higher transpiration rates leading to higher transpiration cooling, in turn, allow maintenance of metabolic processes at a higher rate. Nevertheless, the acclimation for leaf cooling at high temperature comes at the expense of using more water; therefore, utilization of such knowledge for crop improvement will have to come from reasonable balance between these two processes in individual environments.

9.2.2 Plant Processes Underlying Transpiration Responsiveness: Hydraulics and Aquaporin Expression

There is emerging evidence documenting the plant ability to limit canopy TR under elevated VPD is a consequence of limited hydraulic conductivity in its tissues. The limitation of hydraulic conductivity in sorghum plants was shown in the experiments where genotypes contrasting for TR response to high VPD also contrasted for their hydraulic conductivity at the whole-plant level (Choudhary et al. 2013b; Choudhary and Sinclair 2014a). However, the research is ongoing to identify the site (root/leaf/stem) of the major hydraulic restriction in sorghum. Ocheltree et al. (2013) and Choudhary et al. (2013b) studied two sorghum genotypes (SC1205 and SC15) contrasting for TR responsiveness to VPD. Ocheltree et al. (2013) found that these genotypes differed in their leaf hydraulics resistance and that leaf resistance outside the xylem limited whole leaf stomata conductance.

Choudhary et al. (2013b) found substantial variation in aquaporin expression in the sorghum leaves and root while treating the two contrasting genotypes studied previously with aquaporin inhibitors. These results were confirmed in other sorghum germplasm in which expression of the limited-transpiration response differed among genotypes in sensitivity to AQP inhibitors and expression of root AQPs (Choudhary and Sinclair 2014b; Aparna et al. 2017 under review). Genome-wide identification and characterization of aquaporin gene family was recently done in sorghum, and this should provide further clues to understand the role of aquaporins in plant regulation of transpiration processes (Reddy et al. 2015).

References

Aparna K, Anderberg H, Medina S, Kholova J, Chaumont F, Vadez V (2017) Transpiration sensitivity to high vapor pressure deficit increases transpiration efficiency and yield, and is linked to differences in aquaporin gene expression in sorghum (*Sorghum bicolor* (L.) Moench). unpublished

Bunce JA (2003) Effect of water vapor pressure difference on leaf gas exchange in potato and sorghum at ambient and elevated carbon dioxide under field conditions. Field Crop Res 82:37–47

Casa AM, Pressoir G, Brown PJ, Mitchell SE, Rooney WL, Tuinstra MR, Franks CD, Kresovich S (2008) Community resources and strategies for association mapping in sorghum. Crop Sci 48:30–40

Choudhary S, Sinclair TR (2014a) Hydraulic conductance differences among sorghum genotypes to explain variation in restricted transpiration rates. Funct Plant Biol 41:270–275

Choudhary S, Sinclair TR (2014b) Chemical screening for limited-transpiration-rate trait among sorghum genotypes. J Crop Improv 28:377–389

Choudhary S, Mutava RN, Shekoofa A, Sinclair TR, Vara Prasad PV (2013a) Is stay-green trait in sorghum a result of transpiration sensitivity to either soil drying or vapor pressure deficit? Crop Sci 53:1–6

Choudhary S, Sinclair TR, Vara Prasad PV (2013b) Hydraulic conductance of intact plants of two contrasting sorghum lines, SC15 and SC1205. Funct Plant Biol 40:730–738

Gholipoor M, Prasad PVV, Mutava RN, Sinclair TR (2010) Genetic variability of transpiration response to vapor pressure deficit among sorghum genotypes. Field Crop Res 119:85–90

Gholipoor M, Sinclair TR, Prasad PVV (2012) Genotypic variation within sorghum for transpiration response to drying soil. Plant Soil 357:35–40

Jones HG, Corlett JE (1992) Current topics in drought physiology. J Agric Sci 119:291–296

Jordan DR, Hunt CH, Cruickshank AW, Borrell AK, Henzell RG (2012) The relationship between stay-green trait and grain yield in elite sorghum hybrids grown in a range of environments. Crop Sci 52:1153–1161

Kassahun B, Bidinger FR, Hash CT, Kuruvinashetti MS (2010) Stay-green expression in early generation sorghum (*Sorghum bicolor* (L.) Moench) QTL introgression lines. Euphytica 172:351–362

Kholová J, McLean G, Vadez V, Craufurd P, Hammer GL (2013) Drought stress characterization of post-rainy season (rabi) sorghum in India. Field Crop Res 141:38–46

Kholová J, Murugesan T, Kaliamoorthy S, Malayee S, Baddam R, Hammer GL, McLean G, Deshpande S, Hash CT, Craufurd PQ, Vadez V (2014) Modelling the effect of plant water use traits on yield and stay-green expression in sorghum. Funct Plant Biol 41:1019–1034

Ocheltree TW, Nippert JB, Kirkham MB, Prasad PVV (2013) Partitioning hydraulic resistance in grass leaves reveals unique correlations to stomatal conductance during drought. Funct Plant Biol 41:25–36

Parent B, Turc O, Gibon Y, Stitt M, Tardieu F (2010) Modelling temperature-compensated physiological rates, based on the co-ordination of responses to temperature of developmental processes. J Exp Bot 61:2057–2069

Reddy PS, Bhadra Rao SR, Sharma KK, Vadez V (2015) Genome-wide identification and characterization of the aquaporin gene family in *Sorghum bicolor* (L.). Plant Gene 1:18–28

Riar MK, Sinclair TR, Vara Prasad PV (2015) Persistence of limited-transpiration-rate trait in sorghum at high temperature. Environ Exp Bot 115:58–62

Shekoofa A, Balota M, Sinclair TR (2014) Limited-transpiration trait evaluated in growth chamber and field for sorghum genotypes. Environ Exp Bot 99:175–179

Sinclair TR, Hammer GL, Oosterom van EJ (2005) Potential yield and water-use efficiency benefits in sorghum from limited maximum transpiration rate. Funct Plant Biol 32:945–952

Turner NC (1997) Further progress in crop water relations. Adv Agron 58:293–338

Woledge J, Bunce JA, Veronica T (1989) The effect of air humidity on photosynthesis of ryegrass and white clover at three temperatures. Ann Bot 63:271–279

Chapter 10
Pearl Millet

Murugesan Tharanya, Jana Kholová, Kaliamoorthy Sivasakthi, Thiyagarajan Thirunalasundari, and Vincent Vadez

Pearl millet (*Pennisetum glaucum* L. R. Br.) is a well-adapted crop to dry conditions, and it is grown in arid and semiarid regions of Africa and Asia. The average yield of millet for the past 10 years was 0.88 tonnes/ha (FAO, 2016). It is used for grain and fodder and it is a source of micronutrients such as iron and zinc. Though the use of coarse grains like pearl millet declined due to change in pattern of food consumption, their lower prices stimulated the demand of alternate uses such as livestock and poultry feed (Reddy et al. 2013). It can be grown in adverse environments with water deficits and high temperature where generally most other crops fail (Vadez et al. 2012). This makes pearl millet an important crop in the arid and semiarid regions.

Though pearl millet is well adapted to dry conditions, water limitation remains a major constraint to its production (Mahalakshmi et al. 1987). To improve the crop production under drought, plant physiological mechanisms underlying drought adaptation need to be investigated. An important component of achieving yield increase is to understand and modify the pattern of water use through the growing season (Kholova et al. 2013). In this chapter, soil water conservation with soil drying and sensitivity to vapor pressure deficit by pearl millet are reviewed.

M. Tharanya • K. Sivasakthi
International Crops Research Institute for the Semi-Arid Tropics (ICRISAT), Greater Hyderabad, Telangana, India

Bharathidasan University, Tiruchirappalli 620 024, Tamilnadu, India

J. Kholová • V. Vadez (✉)
International Crops Research Institute for the Semi-Arid Tropics (ICRISAT), Greater Hyderabad, Telangana, India
e-mail: v.vadez@cgiar.org

T. Thirunalasundari
Bharathidasan University, Tiruchirappalli 620 024, Tamilnadu, India

T.R. Sinclair (ed.), *Water-Conservation Traits to Increase Crop Yields in Water-deficit Environments*, SpringerBriefs in Environmental Science,
DOI 10.1007/978-3-319-56321-3_10

10.1 Water Conservation with Soil Drying

The earliest results on pearl millet about the relationship between transpiration rate (TR) and volumetric soil water content were presented by Kholova et al. (2010a). In this, assessment of two pearl millet parental genotypes (PRLT 2/89-33 and H77/833-2), which are parents of a mapping population from which a QTL for terminal drought-stress tolerance, i.e., a higher grain yield and a higher panicle harvest index (PNHI) under terminal water-deficit stress, was identified. With gradual soil drying, TR of PRLT 2/89-33 declined at lower fractions of transpirable soil water (FTSW = 0.30) when compared to H77/833-2 (FTSW = 0.49, Fig. 10.1).

The results with these two genotypes were counterintuitive as it was expected that the tolerant genotype would have a transpiration decline at higher soil moisture level, allowing soil water conservation so that water would be available to support reproduction and grain filling. A confounding difference between the two genotypes was that drought-tolerant PRLT 2/89-33 had a lower TR over the entire day than drought-sensitive H77/833-2. The same difference in TR was also observed for detached leaves collected from the field. Kholova et al. (2010a) suggested that the lower TR and slower decrease in soil water content might have delayed the time for the initiation of TR decline. A similar conclusion was drawn in comparing sorghum genotypes (Gholipoor et al. 2012).

The QTL introgression lines were developed in the background of H77/833-2 with PRLT 2/89-33 as the donor parent (Yadav et al. 2002). The drought tolerance

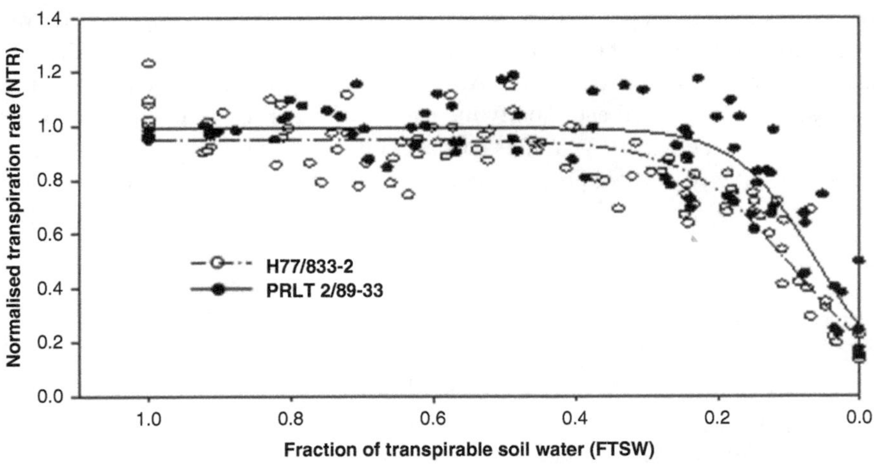

Fig. 10.1 Relationship between the normalized transpiration rate (NTR) and fraction of transpirable soil water (FTSW) of H77/833-2 (*open circle*) and PRLT 2/89-33 (*closed circle*) during vegetative stage. The FTSW thresholds where the transpiration initiated its decline were calculated with plateau regression procedure from SAS. Then the regression lines of the relationships between the NTR and FTSW were drawn by fitting NTR to FTSW data above and below the respective threshold for transpiration decline in each genotype (Kholova et al. 2010a)

QTL followed a similar pattern to that of parents, i.e., FTSW of low TR introgression lines was similar to PRLT 2/89-33 and the high TR introgression lines had FTSW similar to H77/833-2.

10.2 Water Conservation from Sensitivity to Vapor Pressure Deficit

The report on genotypic difference in TR of pearl millet by Kholova et al. (2010a) was extended to examining response of TR sensitivity to atmospheric vapor pressure deficit (VPD). It was hypothesized that one of the mechanisms of water-deficit adaptation would be conservation of soil water under elevated VPD conditions including under fully irrigated conditions.

Kholova et al. (2010b) studied two pairs of pearl millet genotypes, PRLT 2/89-33 and H77/833-2, and ICMB 841-P3 and 863B-P2, that had been previously shown to contrast in drought tolerance under field conditions (Yadav et al. 2002; Serraj et al. 2005). H77/833-2 and ICMB 841-P3 were drought-sensitive genotypes, whereas PRLT 2/89-33 and 863B-P2 were drought-tolerant genotypes. TR of these genotypes was measured every hour in both controlled chamber and in outdoor conditions as VPD varied (0.6 to 2.8 kPa). In the controlled-environment chamber under low VPD conditions (around 1 kPa), there was small but significant genotypic difference. But as the VPD increased, the genotypic difference in TR increased (Fig. 10.2).

Under natural outdoor conditions, the pattern of TR closely followed the daily pattern of VPD. Genotypic differences were not observed under low VPD (less than 3 kPa; until 10:00 standard time). At VPD greater than 3 kPa, the increase in TR was lower in PRLT 2/89-33 than H77/833-2, and the largest genotypic difference was found at high VPD (at the time between 11:00 and 15:00 standard time) (Kholova et al. 2010b).

In addition to the lower TR of PRLT 2/89-33, the TR increase with VPD showed a breakpoint or restriction at 1.91 kPa (Kholova et al. 2010b) in the results from the growth chamber study (Fig. 10.2). Similarly, the other drought-tolerant genotype 863B-P2 had a breakpoint at 1.75 kPa. In contrast, drought-sensitive genotypes H77/833-2 and ICMB 841-P3 had linear increases in TR with increasing VPD across the whole range of VPD. Similar measurements were made also in the QTL introgression lines. Among these, the water-deficit adapted lines had TR response to VPD similar to PRLT 2/89-33 and 863B-P2, and water-deficit sensitive lines showed a TR response to VPD similar to H77/833-2 and ICMB 841-P3 in most cases.

The TR response to increasing VPD was also compared between soil and hydroponics growth conditions (Tharanya et al. 2017 under review). In both growing systems, the VPD-sensitive genotype PRLT 2/89-33 had lower TR at all VPD than the VPD-insensitive genotype H77/833-2 which confirmed the earlier results (Kholova et al. 2010b). Hence, the TR difference in response to VPD between the genotypes was consistent across both hydroponics and soil-based growth systems. However, in the

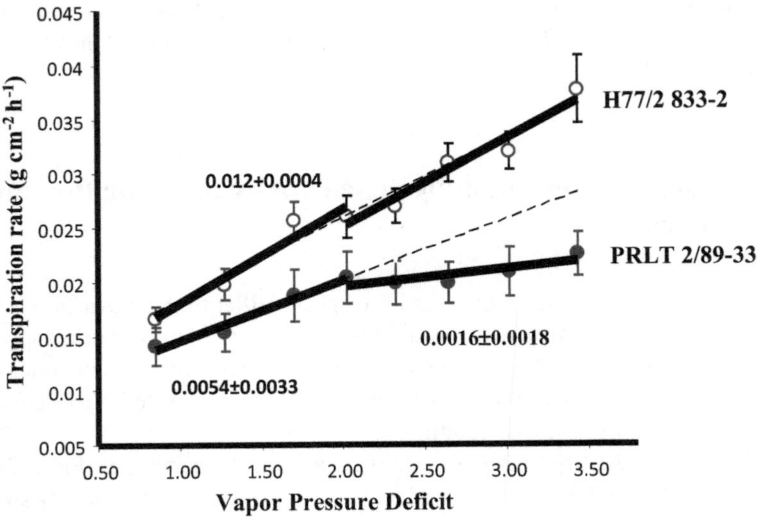

Fig. 10.2 Transpiration rate (TR, gcm^{-2} h^{-1}) under well-watered conditions of H77/833-2 (*open circle*) and PRLT 2/89-33 (*closed circle*) exposed to increased VPD regimes in a controlled-environment chamber. The plants were tested at vegetative stage and the values are the means (±SE) of six replications. Numbers above/below the regression curves are the slopes (±SE) of the TR response. The breakpoint of the slope was 1.91 ± 0.75 kPa for PRLT 2/89-33 and the breakpoint of the slope was not significant for H77/833-2 (Kholova et al. 2010b)

hydroponic system, the breakpoint response which was observed in soil earlier was not found. It was suggested that this response may have resulted because the hydroponically grown plants may not have developed complete apoplastic barriers, i.e., Casparian bands in the hypodermis and endodermis of the roots (Peterson 1998; Zimmerman and Steudle 1998; Freundl et al. 2000) resulting in less limitation to water flow.

10.3 Water Transport Pathways

The restricted TR response of PRLT 2/89-33 observed at VPD greater than 2 kPa (Kholova et al. 2010b) occurred rapidly in the 15 min of transitions between successive, increasing VPD. The fact that the shift from one linear response slope to another occurred in such a short time implies the basis for the response was more likely to be mediated by hydraulic signals rather than by biochemical mediators (Kholova et al. 2010b). Hence, it was hypothesized that the genotypic difference in transpiration response to VPD might be related to the difference in channeling of water in the water transport pathways.

To explore this hypothesis, an initial experiment compared TR response to increasing VPD of whole plants and de-rooted shoots. The initial slope of TR versus VPD reflecting plant conductivity was lower in whole plants than in de-rooted shoot. This result indicated there was resistance to water transport in roots that was removed in the de-rooted plants (Tharanya et al. 2017 under review).

Further evidence of the importance of root hydraulic conductivity was found in a genotypic comparison of the initiation of TR decrease of detached leaves under high VPD (2.67 kPa) and low VPD (1.27 kPa). The experiment with detached leaves did not show significant difference in the TR for drought-sensitive H77/833-2 and for drought-tolerant PRLT 2/89-33, even though these genotypes showed different response in TR of intact plants at high VPD. These results indicated that roots hydraulics limited TR under high VPD in these pearl millet lines.

According to the composite water transport model in roots, there are three possible water transport pathways – apoplastic, symplast, and transcellular paths (Steudle and Peterson 1998). In the apoplastic pathway, the water moves through the intercellular spaces around the protoplast driven by a hydrostatic pressure gradient. In the symplast pathway, water moves through the plasmodesmata, and in transcellular path, water crosses two membranes per cell. Experimentally symplast and transcellular pathways were inseparable and they are termed together as "cell-to-cell pathway" (Steudle and Peterson 1998). Transcellular transport appears to involve specialized membrane protein transporters called aquaporin (AQP), which can regulate AQP-mediated water transport (Lee and Zwiazek 2015). In pearl millet, it was found that both apoplastic and transcellular pathways contributed to water transport (Tharanya et al. 2015), although there were genotypic differences in the contribution of each pathway. Under low VPD (1.8 kPa), the transpiration in intact plants of VPD-insensitive H77/833-2 was decreased more by treatment with the AQP inhibitor H_2O_2 than in VPD sensitive-genotype PRLT 2/89-33 (Fig. 10.3). Similarly, TR of detached leaves from H77/833-2 decreased more than in PRLT 2/89-33 when treated with

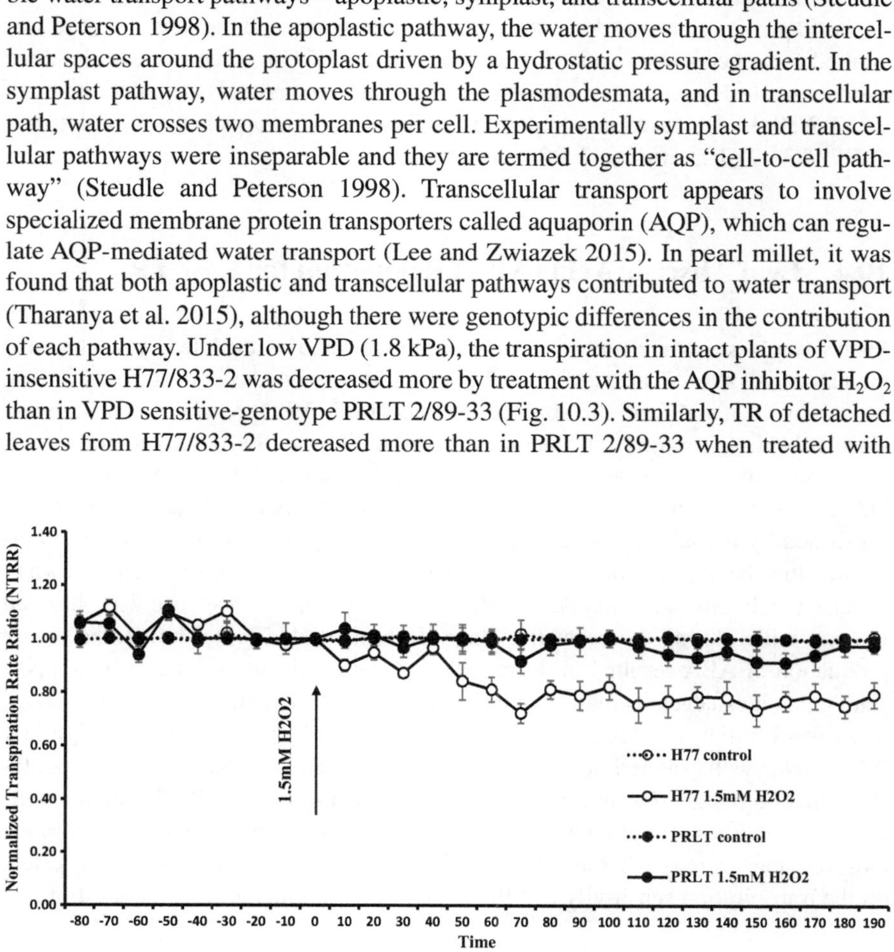

Fig. 10.3 Transpiration decline (double normalized values, NTRR) of H77/833-2 (*open circle*) and PRLT 2/89-33 (*closed circle*) after exposure AQP inhibition (1.5 mM H_2O_2) at constant low VPD (1.8 kPa). The *dotted lines* represent the NTRR of untreated control plants. Each point is the mean of NTRR (n = 5) and the *error bars* are the SE of the mean. Experiment was done with hydroponic plants and tested in growth chamber at vegetative stage. The timings with negative symbols refer prior to inhibitor application and the timings with positive symbols refer after inhibitor application

$AgNO_3$ (Tharanya et al. 2017 under review). These results indicated that the trans-membrane pathway was more important in H77/833-2 than in PRLT 2/89-33.

The complimentary conclusion that the apoplastic pathway was more important in PRLT 2/89-33 was tested by making direct measurements of root hydraulic conductivity. Root hydraulic conductivity was measured by placing the root system in a Scholander pressure chamber, sealing the plant at the hypocotyl level, and removing the shoot by a single cut. The remaining root system was subjected to a series of pressures (0.1, 0.2, and 0.3 MPa) for 15 min each, and the exuded sap was collected every 5 min at the cut surface using a pre- weighed Eppendorf stuffed with tissue paper. Sap collection was done thrice for each pressure, and the next pressure was applied once a constant exudation rate was reached. The average value of the three exudation rate samples was normalized with root surface area. The exudation rate of PRLT 2/89-33 decreased more upon the application of an apoplastic inhibitor than in H77/833-2 (Tharanya et al. 2017 under review).

10.4 Leaf Abscisic Acid (ABA) Regulation for Low TR

Plant water losses as a result of transpiration are known to be regulated by biochemical mediators like abscisic acid (ABA) and/or by hydraulic signals (Comstock 2002). The evaporative demand from the atmosphere (increase in VPD) results in leaf stomatal closure that has proposed to be mediated by ABA (Grantz 1990). ABA is produced in roots and transported to the shoot and then to the leaf stomatal guard cells (Zhang and Davies 1991; Tardieu et al. 2010). The involvement of ABA in stomatal closure was confirmed by partial root-drying experiments (Gowing et al. 1990). Another report showed that the response of partial stomatal closure by ABA occurs in plants grown in adverse conditions like water deficit (Raschke 1975). Similar response was observed in pearl millet during the vegetative stage (Henson et al. 1983). Though the higher production of ABA resulted in decreased stomatal conductance, there was a report stating that it can even increase root hydraulic conductivity (Thompson et al. 2007).

In pearl millet, Kholova et al. (2010b) found that PRLT 2/89-33 had high leaf ABA under well-watered conditions during the vegetative stage. It was hypothesized that high leaf ABA in a VPD-sensitive genotype (PRLT 2/89-33) under well-watered conditions might favor the reduction in water loss by partial stomatal closure under increased TR at elevated VPD. However, whether ABA has any effect on the transpiration sensitivity to VPD in pearl millet remains an open question.

10.5 Gene Expression of Water-Saving Traits

Two independent mapping populations of recombinant inbred lines (RILs) contrasting for water-deficit adaptation were developed from the cross between H77/833-2 and PRLT 2/89-33 and the other from the cross between ICMB 841(water-deficit sensitive) and ICMB 863 (water-deficit adapted) (Hash et al. 1999; Yadav et al. 2004). A major terminal water-deficit adapted QTL was identified in those two inbred mapping populations (Yadav et al. 2002, 2004; Bidinger et al. 2007).

Kholova et al. (2012) mapped the water conservation traits in the RIL population developed from H77/833-2 and PRLT 2/89-33. Three QTLs explaining 13–26% of phenotypic variation in TR were identified on LG2. These QTLs associated with water-saving traits were reported to co-map with previously identified major terminal water-deficit adapted QTL on LG2, adding evidence for a close relationship between this terminal water-stress tolerance QTL and traits controlling plant water use.

Kakkera et al. (2015) did similar work in a RIL population developed from the mating of ICMB 841 and ICMB 863. Four QTLs for TR from ICMB 841 were reported to be associated with increased TR under high VPD. The major QTL out of these four, explaining 36% of the phenotypic variation, was mapped on LG6. The QTL for low TR from ICMB 863 was reported to co-map with yield components under water-deficit stress on LG6 indicating the existence of a relationship between low TR and yield components under water-deficit conditions. However, in this case no co-mapping was found with the earlier QTL for terminal drought tolerance in LG2 in that population (Yadav et al. 2003).

Since higher hydraulic conductivity of the root is closely linked with increase in expression of several AQP genes (e.g., Knipfer et al. 2011; Ehlert et al. 2009; Hachez and Chaumont 2010), an attempt was made to explore the AQP gene expression in pearl millet. Reddy et al. (in preparation) cloned AQP genes (PgPIP1;1, PgPIP1;2, PgPIP2;1, PgPIP2;3, PgPIP2;6, PgTIP1;1, and PgTIP2;2) using the homology-based cloning approach. Two pairs of pearl millet high resolution cross (HRC) materials contrasting for the sensitivity of TR response to increasing VPD were used in this study. Based on phylogenetic analysis, the cloned pearl millet PgAQP genes appeared to be similar to maize AQPs.

During the early morning hours under low VPD (1.2 kPa), PgAQP genes had a slightly higher expression in VPD-insensitive genotypes than the VPD-sensitive genotypes. During the afternoon hours of the day, the expression of AQPs generally decreased regardless of the VPD conditions. The expression of AQP genes from root and leaf tissues of VPD-sensitive genotypes had a lower transcript abundance than the VPD-insensitive genotypes under high VPD conditions (4.21 kPa). Interestingly, among the PIP2 subfamily, PIP2.1, PIP2.3, and PIP2.6 showed an increase in transcript abundance under high VPD conditions. This could provide an explanation for the lower TR of the VPD-sensitive genotypes under high VPD conditions, in comparison to the VPD-insensitive genotypes.

In summary, the QTL mapping work done so far has revealed a close involvement of these water-saving traits in the terminal water-deficit adaptation QTLs. The mapped QTLs for water-saving traits might be helpful in designing environment-specific ideotypes which would cater the need for water-limited conditions. Differences in VPD sensitivity appeared to correlate with differences in the expression of PgAQP genes in the roots under high VPD.

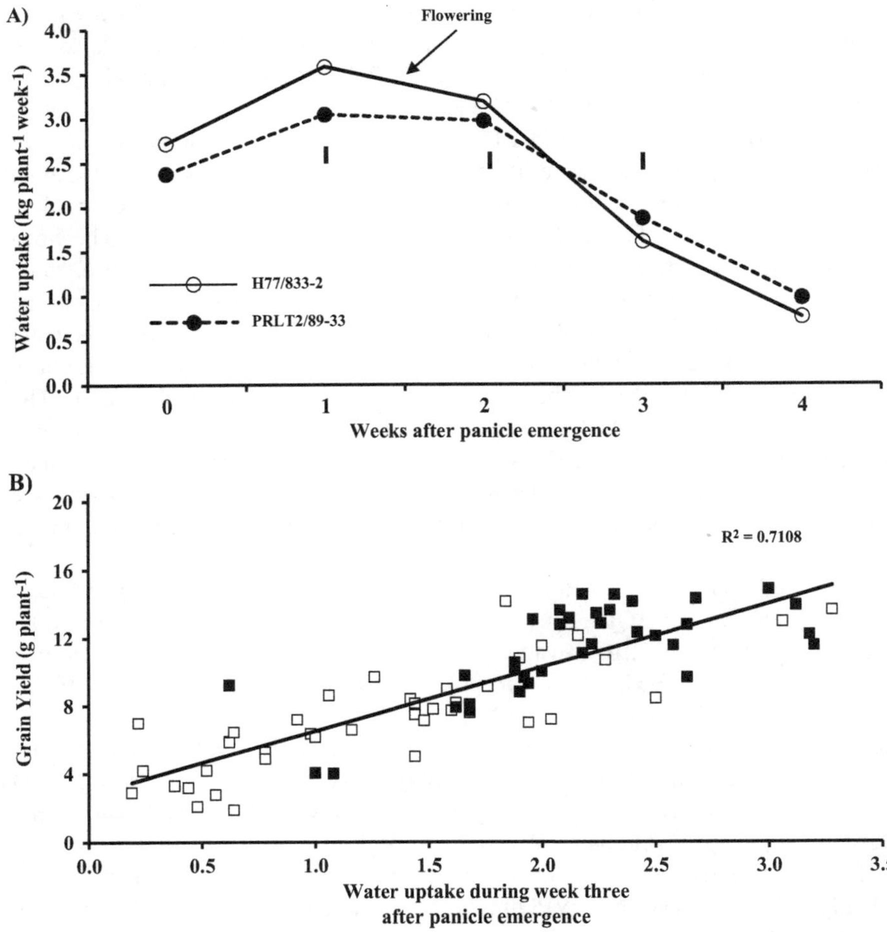

Fig. 10.4 (**a**) Pattern of water extraction from the time of panicle emergence until maturity in H77/833-2 (*open circle*) and PRLT 2/89-33 (*closed circle*). Water uptake (kg $plant^{-1}$ $week^{-1}$) was measured under water stress and the vertical bars indicate LSD at $P < 0.05$. The absence of bars above specific dates indicates that genotypic differences were not significant. The timing of intervals indicates the number of weeks after panicle emergence, and the water uptake data assigned to each date corresponds to the time period (usually 1 week) preceding that date. The data are the means of 20 lysimeters exposed to early water-deficit stress. (**b**) Relationships between the total grain yield at maturity and the water extracted (kg $plant^{-1}$) during the third week after panicle emergence under early stress (open and closed squares) conditions. Lysimeters with an initial full soil profile are represented by closed symbols, while those with an initial unfilled profile are represented by open symbols (Vadez et al. 2013)

10.6 Effects of Water Conservation on Yield

To explore water conservation mechanism in pearl millet in experimental conditions approximating field conditions, the water extraction pattern was followed in a lysimeter system providing plants the same volume and space available as in the field (Vadez et al. 2013). A terminal water-deficit stress treatment was applied by stopping irrigation at booting. The water-deficit adapted genotype (PRLT 2/89-33) extracted less water before flowering, than in the absence of stress, and so the water conserved before flowering was available for use after flowering during the grain-filling period. In contrast, the water-deficit sensitive genotype H77/833-2 extracted more water before flowering and ran short of water during grain filling (Fig. 10.4a). The temporal difference between genotypes in the seasonal pattern of water extraction provided more water for better grain filling and yield in PRLT 2/89-33 than observed in H77/833-2. A clear positive relationship was found between water uptake in week three after panicle emergence, when irrigation was stopped, and grain yield (Fig. 10.4b). It was hypothesized that PRLT 2/89-33 was able to restrict transpiration under high VPD in this summer trial, in agreement with similar evidence in these lines reported earlier (Kholova et al. 2010a, b), which led to water conservation. Similar results in outdoors conditions were confirmed in the LeasyScan setup (Vadez et al. 2015), where indeed PRLT 2/89-33 displayed a systematic transpiration restriction in all days when VPD was above 2.5 kPa.

References

Bidinger FR, Nepolean T, Hash CT et al (2007) Identification of QTLs for grain yield of pearl millet [*Pennisetum glaucum* (L.) R. Br.] in environments with variable moisture during grain filling. Crop Sci 47(3):969–980

Comstock JP (2002) Hydraulic and chemical signaling in the control of stomatal conductance and transpiration. J Exp Bot 53:195–200

Ehlert C, Maurel C, Tardieu F et al (2009) Aquaporin-mediated reduction in maize root hydraulic conductivity impacts cell Turgor and leaf elongation even without changing transpiration. J Plant Physiol 150:1093–1104

FAO database (2016) Available online http://faostat3.fao.org/download/Q/QC/E

Freundl E, Steudle E, Hartaung W (2000) Apoplastic transport of abscisic acid through of maize: effect of exodermis. Planta 210:222–231

Gholipoor M, Sinclair TR, Prasad PVV (2012) Genotypic variation within sorghum for transpiration response to drying soil. Plant Soil 357:35–40

Gowing DJG, Davies WJ, Jones HG (1990) A positive root-sourced signal as an indicator of soil drying in apple (*Malus domestica* Borkh). J Exp Bot 41:1535–1540

Grantz DA (1990) Plant response to atmospheric humidity. Plant Cell Environ 13:667–679

Hachez C, Chaumont F (2010) Aquaporins: a family of highly regulated multifunctional channels. Adv Exp Med Biol 679:1–18

Hash CT, Yadav RS, Cavan GP et al (1999) Marker-assisted backcrossing to improve terminal drought tolerance in pearl millet. In: Proceedings of a Strategic Planning Workshop on "Molecular Approaches for the Genetic Improvement of Cereals for Stable Production in

Water-limited Environments", June 21–25, CIMMYT, El Batan, Mexico (Ribaut J-M and Poland D. Eds.) CIMMYT Mexico, D.F. Mexico. pp 114–11

Henson E, Mahalakshmi V, Alagarswamy G et al (1983) An association between flowering and reduced stomatal sensitivity to water stress in pearl millet [*Pennisetum americanum* (L.) Leeke]. Ann Bot 52:641–648

Kakkera A, Nepolean T, Srivastsava RK et al (2015) Quantitative trait loci associated with constitutive traits controlling water use in pearl millet [*Pennisetum glaucum* (L.) R. Br.] Plant Biol 17(5):1073–1084. doi:10.1111/plb.12343

Kholová J, Hash CT, Kakkera A et al (2010a) Constitutive water conserving mechanisms are correlated with the terminal drought tolerance of pearl millet [*Pennisetum glaucum (L.) R.Br.*] J Exp Bot 61:369–377

Kholová J, Hash CT, Kumar PL et al (2010b) Terminal drought –tolerant pearl millet [*Pennisetum glaucum* (L.) R.Br.] have high leaf ABA and limit transpiration at high vapor pressure deficit. J Exp Bot 61:1431–1440

Kholová J, Nepolean T, Hash CT et al (2012) Water saving traits co-map with a major terminal drought tolerance quantitative trait locus in pearl millet [*Pennisetum glaucum* (L.) R. Br.] Mol Breed 30:1337–1353

Kholová J, McLean G, Vadez V et al (2013) Drought stress characterization of post-rainy season (rabi) sorghum in India. Field Crop Res 141:38–46

Knipfer T, Basse M, Verdeil JL et al (2011) Aquaporin – facilitated water uptake in barley (*Hordeum vulagare* L.) roots. J Exp Bot 62:4115–4126. doi:10.1093/jxb/err075

Lee SH, Zwiazek JJ (2015) Regulation of aquaporin-mediated water transport in Arabidopsis roots exposed to NaCl. Plant Cell Physiol 4:750–758

Mahalakshmi V, Bidinger FR, Raju DS (1987) Effect of timing of water deficit on pearl millet (*Pennisetum americanum*). Field Crop Res 15:327–339

Peterson CA (1998) Exodermal casparian bands, their significance for ion uptake by roots. Plant Physiol 72:204–208

Raschke K (1975) Stomatal action. Annu Rev Plant Physiol 26:309–340

Reddy AA, Yadav OP, Malik DP et al (2013) Utilization pattern, demand and supply of pearl millet grain and fodder in Western India. International Crops Research Institute for the semi-Arid tropics. Working Paper Series no. 37. Available at: http://oar.icrisat.org/id/eprint/6694

Reddy PS, Tharanya M, Sivasakthi K et al Molecular Cloning and Expression Analysis of Aquaporin Genes in Pearl Millet [*Pennisetum glaucum* (L) R. Br.] genotypes contrasting in their Transpiration Response to High Vapour Pressure Deficits. (Manuscript under preparation)

Serraj R, Hash CT, Rizvi SMH et al (2005) Recent advances in marker assisted selection for drought tolerance in pearl millet. Plant Prot Sci 8:334–337

Steudle E, Peterson CA (1998) How does water get through roots? J Exp Bot 322:775–788

Tardieu F, Parent B, Simonneau T (2010) Control of leaf growth by abscisic acid: hydraulic or non-hydraulic process? Plant Cell Environ 33:636–647

Tharanya M, Sivasakthi K, Barzana G et al (2015) Pearl millet (*Pennisetum glaucum*) genotypes contrasting for transpiration sensitivity also differ in their dependence on Aquaporin mediated and apoplastic water transport pathway. In: Tropical Agriculture Conference, 16–18 November, Brisbane, Australia. 2015. Book of Abstracts

Tharanya M, Sivasakthi K, Barzana G, Kholová J, Thirunalasundari T, Vadez V (2016) Pearl millet [*Pennisetum glaucum (L.) R. Br.*] contrasting for the transpiration response to vapour pressure deficit also differ in their dependence on the symplastic and apoplastic water transport pathways. Funct Plant Biol (under review)

Thompson AJ, Andrews J, Mulholland BJ et al (2007) Overproduction of abscisic acid in tomato increases transpiration efficiency and root hydraulic conductivity and influences leaf expansion. Plant Physiol 143:1905–1917

Vadez V, Hash CT, Bidinger FR et al (2012) Phenotyping pearl millet for adaptation to drought. Front Physiol 3:386. doi:10.3389/fphys.2012.00386

Vadez V, Kholová J, Yadav RS et al (2013) Small temporal differences in water uptake among varieties of pearl millet (*Pennisetum glaucum (L.) R. Br.)* are critical for grain yield under terminal drought. Plant Soil 371(1):447–462. doi:10.1007/s11104.013.1706.0

Vadez V, Kholova J, Hummel G et al (2015) LeasyScan: a novel concept combining 3D imaging and lysimetry for high-throughput phenotyping of traits controlling water budget. J Exp Bot. doi:10.1093/jxb/erv251

Yadav HCT, Bidinger FR et al (2002) Quantitative trait loci associated with traits determining grain and stover yield in pearl millet under terminal drought stress. Theor Appl Genet 104:67–83

Yadav RS, Bidinger FR, Hash CT et al (2003) Mapping and characterization of QTL × E interactions for traits determining grain and stover yield in pearl millet. Theor Appl Genet 106:512–520

Yadav RS, Hash CT, Bidinger FR et al (2004) Genomic regions associated with grain yield and aspects of post-flowering drought tolerance in pearl millet across stress environments and testers background. Euphytica 136:265–277

Zhang J, Davies WJ (1991) Antitranspirant activity in the xylem sap of maize plants. J Exp Bot 42:317–321. doi:10.1093/jxb/42.3.317

Zimmerman HM, Steudle E (1998) Apoplastic transport across young maize roots: effects of the exodermis. Planta 206:7–19

Chapter 11
Wheat

Walid Sadok

Wheat was domesticated over 8000 years ago in the Fertile Crescent, and its socio-economic and cultural importance has continued to increase. Currently, wheat is the most widely grown and traded crop and the second most produced commodity after rice, at 749 million tons (FAO 2016). Wheat is currently a major staple food to approximately one fifth of humanity (FAO 2011). A major national goal in many countries is an adequate production of wheat since wheat grain shortages have been linked to social and political turmoil. In several North African and Middle Eastern countries, recent shortages of grain have resulted in societal revolts, sometimes triggering regime change (Walton and Seddon 2008; Nouri 2013).

Worldwide, the single, most important factor that jeopardizes wheat food security is drought. While wheat is grown under a variety of environments (250–1750 mm of annual precipitation), over 60% of wheat acreage is rain fed (Curtis 2002; Rosegrant et al. 2002), which makes its production particularly prone to variability in precipitation. As a result, it is currently estimated that more than 50% of wheat production environments in all continents are prone to regular, yield-reducing water-deficit events (Farooq et al. 2014). Developing superior, drought-resilient cultivars that maintain productivity under water-limited environments should deliver the type of solution needed to address at least in part wheat food security.

W. Sadok (✉)
Department of Agronomy and Plant Genetics, University of Minnesota,
411 Borlaug Hall, 1991 Upper Buford Circle, St. Paul, MN 55108, USA
e-mail: msadok@umn.edu

T.R. Sinclair (ed.), *Water-Conservation Traits to Increase Crop Yields in Water-deficit Environments*, SpringerBriefs in Environmental Science,
DOI 10.1007/978-3-319-56321-3_11

11.1 Water Conservation in Response to Progressive Soil Drying

Over the last three decades, enhancing wheat drought tolerance, particularly in chronically challenged environments such as the Mediterranean basin and the Australian wheat belt, has developed into a major, long-term research goal. The focus of such research was on observations on such factors as early vigor, water use efficiency, carbon isotope discrimination, leaf conductance, canopy temperature, and enhanced water acquisition, most of which contributed to the release of drought-tolerant cultivars (for details see Farquhar and Richards 1984; Condon et al. 1990, 2002; Fischer et al. 1998; Rebetzke and Richards 1999; Rebetzke et al. 2003, 2013; Wasson et al. 2012). Despite this progress, there is little evidence that water conservation traits, particularly during progressive soil drying, have received extensive attention.

Recently, Schoppach and Sadok (2012) examined the genetic variability in transpiration response to decreasing fractions of transpirable soil water (FTSW) among a group of eight Australian cultivars and breeding lines that included one drought-sensitive cultivar (Kukri). Despite the relatively limited number of genotypes, the results showed significant variability in the FTSW threshold at which transpiration started to decrease, with values ranging from 0.43 to 0.52. The highest FTSW threshold of 0.52 was observed for the highly drought-tolerant breeding line RAC875, which indicated that this genotype may owe its drought tolerance capability to an enhanced sensitivity to soil drying. Such a hypothesis is consistent with the investigation of Izanloo et al. (2008), during which imposition of a cyclic drought revealed a *conservative phenotype* from RAC875 which exhibited a lower stomata conductance during water-deficit stress and ultimately the highest grain yield under water deficit.

Compared to the large variation in FTSW threshold reported for peanut or maize (see Chaps. 5 and 8), the threshold range found so far in wheat is relatively limited. Therefore, it is expected that phenotyping a larger and a more diverse genotypic pool would reveal greater variation in the FTSW threshold.

11.2 Water Conservation in Response to Increasing Atmospheric Vapor Pressure Deficit

11.2.1 Diversity in Transpiration Response to VPD in Wheat and Relationship with Yield Performance under Drought

The eight genotypes examined in Schoppach and Sadok (2012) were also characterized for their transpiration rate (TR) response to increasing VPD. Surprisingly, the TR response to increasing VPD was found to vary much more than TR responses to FTSW. Further, the investigation revealed that the drought-tolerant genotype

RAC875 exhibited the lowest slope of TR response to VPD among all studied lines, indicating that the water saving capability of this line stems at least in part from reducing TR loss in response to increasing VPD.

In a follow-up investigation, Schoppach et al. (2017) examined the diversity in TR response curves to increasing VPD within a larger and diverse pool of 23 Australian wheat genotypes. An additional goal of that investigation was to select those genotypes such that their year of release covered a period of 120 years of breeding, ranging from 1890 to 2008. Because Australian wheat yields more than quadrupled during that same period (Kirkegaard and Hunt 2010), this investigation allowed a test of the hypothesis that such yield increases might have been related to changes in a water-conservation trait, namely, TR response to VPD. The experiments were done in a growth chamber with a constant temperature of 31 °C. The vapor pressure deficit (VPD) was increased incrementally to measure TR at each VPD level.

The initial outcome of the investigation by Schoppach et al. (2017) was that all the studied cultivars displayed a bilinear response of TR to VPD. The VPD at which there was a change in slope varied among genotypes between 1.9 and 2.3 kPa (Fig. 11.1). The consistent nature of the bilinear responses indicated that over the 120 years of selection, Australian breeders "unconsciously" favored a water-conservation drought tolerance trait that relied on a change in TR sensitivity to VPD over a relatively stable VPD threshold.

Another intriguing finding of the study of the 23 Australian genotypes was that the slope in TR at VPD greater than the VPD threshold was correlated strongly and positively ($P < 0.0001$; $R^2 = 0.72$) with the year of release, suggesting that over

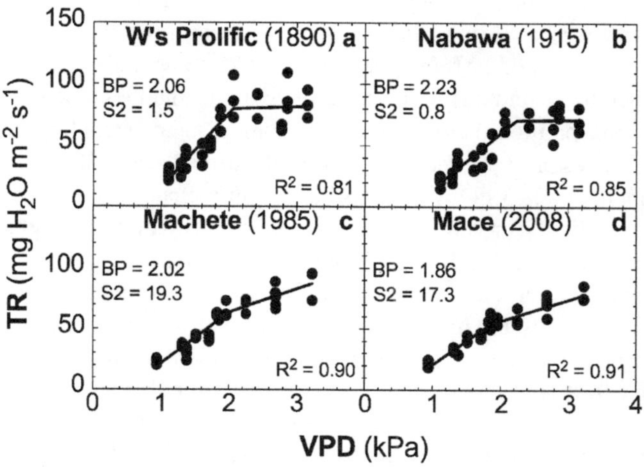

Fig. 11.1 Genetic variability in whole-plant transpiration response (TR) to increasing vapor pressure deficit (VPD) among four representative genotypes from the studied group of 23 cultivars released from 1890 to 2008 (Adapted from Schoppach et al. 2017). In each panel **(a-d)**, the name, year of release, and the values of the breakpoint (BP), Slope 2 (S2) and R^2, of the segmented regressions are indicated

120 years of selection resulted in the progressive increase in canopy conductance under high VPD conditions (Schoppach et al. 2017). This might seem counterintuitive considering that the increase in the TR slope at high VPD should diminish the water-conserving capability of the more modern genotypes, but this might come with the positive trade-off of enhancing the opportunity for CO_2 fixation that comes with increasing canopy conductance. Such a possibility is in line with the fact that the increase of the TR slope at high VPD over time tracks the historic yield increases that took place over the same time period. As with the FTSW findings, further research is needed to clarify the physiological basis of the observed variation and quantify their impacts on grain yield.

11.2.2 Involvement of Roots in Restricted TR in Response to VPD

Research previously carried out on the soybean slow-wilting line PI 416937 conclusively demonstrated the existence of a leaf-based hydraulic restriction to the water flow between the leaf xylem and the guard cells which is responsible for the restricted TR above a VPD threshold and the resulting maximum TR trait (see Chap. 4 for details). In contrast, in wheat, no such involvement was demonstrated so far. Schoppach et al. (2014) hypothesized that similar to whole-plant responses (Sect. 11.2.1), de-rooted shoots of RAC875 would exhibit a lower TR under high VPD than a drought-sensitive cultivar (Kukri). Unexpectedly, several experiments converged to show that the lower TR exhibited by RAC875 were expressed only if the roots were present. The resulting hypothesis of a hydraulic restriction originating from the roots was confirmed in a follow-up experiment that revealed a lower hydraulic conductivity of the root system of RAC875 compared to Kukri (Fig. 11.2a). Taken together, these findings indicate a more complex and less localized control of whole-plant transpiration that might involve long-distance signaling. In fact, those findings are consistent with the report of Kudoyarova et al. (2011) who found in wheat that increases in VPD were associated with a change in root hydraulic conductivity. Interestingly, in that study, changes in root hydraulic conductivity were mediated by abscisic acid (ABA) exported from the shoots.

The mechanism for the changes in variations in hydraulic conductivity may involve aquaporins (AQPs) in the water transport pathway. In wheat, Bramley et al. (2009) presented evidence indicating direct involvement of AQPs in hydraulic conductivity of wheat root cells on the basis of experiments using the AQP blocker mercury (as $HgCl_2$). To test the hypothesis of an involvement of root AQPs in the limited hydraulic conductivity of the roots of RAC875, Schoppach et al. (2014) subjected shoots and whole plants of RAC875 and Kukri to different types of aquaporin inhibitors, including $HgCl_2$ and the mercury scavenger ß-mercaptoethanol. Combined, the experiments supported the hypothesis that the lower hydraulic conductivity of the roots of RAC875 was associated with a lower sensitivity to the

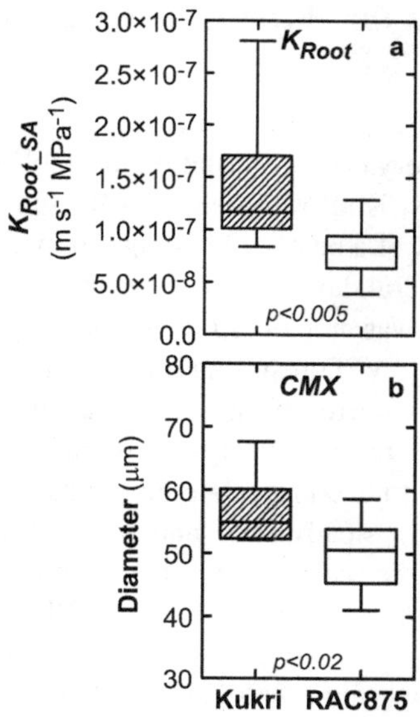

Fig. 11.2 Key differences in root traits between the water saving genotype RAC875 and the check cultivar Kukri (adapted from Schoppach et al. 2014). Panel (**a**): Differential root hydraulic conductivity calculated on a root surface area basis in the roots of 22- to 24-day-old plants. Data is the average of 12 observations. Removing the Kukri outlier datum maintains the statistical difference (*t*-test, *P*-value) significant at 0.008. Panel (**b**): difference in central metaxylem (CMX) diameters in the longest seminal roots of 22-day-old plants. Each *box* and whisker represents a total of 12 observations. When significant, statistical differences are indicated with *P*-values (*t*-tests, ns., non-significant difference). The *box* represents the median and the 25th–75th percentiles and the whiskers represent the largest and smallest values

mercury treatments indicating the lack of a mercury-sensitive aquaporins (AQPs) in this genotype. Because root AQPs are responsive to regulation by hormones such as ABA (Mahdieh and Mostjeran 2009), the inhibitor findings are consistent with the possibility of a stronger involvement of the roots in regulating whole-plant transpiration in wheat.

Another possibility for the mechanistic basis of the variations in hydraulic conductivity is root anatomy. Richards and Passioura (1989) published a 5-year study documenting a wheat breeding program that increased yields up to 11% by reducing the central metaxylem diameter (CMX) of the seminal roots from 65 to less than 55 µm. The guiding hypothesis was that such a restriction in CMX diameters would limit water consumption and therefore enhance water availability during anthesis and yield performance of rain-fed wheat under terminal drought conditions. An examination of the CMX in the seminal roots of the drought-tolerant line RAC875 and the drought-sensitive cultivar Kukri by Schoppach et al. (2014) revealed that the water-conserving line indeed exhibited significantly lower CMX diameters. Interestingly, the average CMX diameter of the seminal roots of RAC875 was well below the limit proposed by Richards and Passioura (1989), being 49.8 ± 1.8 µm (Fig. 11.2b).

11.2.3 Genetic Basis for Water Conservation in Response to VPD

Until recently, there was very little information available on the genetic basis of TR response to VPD in plants. Kholova et al. (2012) used a gravimetric approach to phenotype whole-plant TR of a pearl millet mapping population, but such measurements were made over several days at two points in time, such that constructing TR response curves to VPD was not possible. Phenotyping TR response curves to increasing VPD per se (i.e., VPD resulting only from changes in relative humidity at a stable temperature) is currently difficult at a high throughput level. This is because such experiments have to be typically carried out in growth chambers which are often too small to accommodate hundreds of genotypes simultaneously and incapable of imposing steady-state and large enough variations in relative humidity. As a result, a more realistic approach is to track whole-plant TR change in open environments (greenhouse or field) where VPD is driven by both an increase in temperature and decrease in relative humidity. Schoppach et al. (2016) developed a manual gravimetric phenotyping protocol that relied on such an approach to characterize simultaneously and within a single day the response curves of TR to increasing VPD under greenhouse conditions.

The screen of genotype response to VPD was performed on progeny lines of the mating of the genotypes RAC875 and Kukri. A mapping population of doubled haploid (DH) lines was generated that consisted of approximately 300 genotypes (Fleury et al. 2010). Therefore, this population offered an excellent basis for phenotyping and identifying the genetic basis of TR responses to VPD in wheat. Within this group of lines, a subset of 143 lines were randomly selected among genotypes that flowered within a 3-day interval (Schoppach et al. 2016). These lines were previously characterized in the field for several yield component traits (Bennett et al. 2012; Bonneau et al. 2013).

In the RAC875 × Kukri DH population, six QTL collectively explaining 83.4% of the genetic variance were detected for the slope of TR response to VPD (SLP). The most robust of these (and of all 68 QTL detected for other traits) was a QTL for SLP on chromosome 5A, which individually explained >25% of the genetic variance (LOD = 6.7). The peak region of this QTL was mapped to a group of 12 drought-related genes among which was DRIP1, a negative regulator of DREB2A, a transcription factor involved in drought tolerance (Engels et al. 2013). In addition, in the same investigation, independent RNASeq experiments revealed that 9 of the 12 genes were found to be preferentially expressed in the roots of wheat cv. Chinese Spring independently from the growth stage. Importantly, the annotations of these genes highlighted their involvement in root development, water transport, water deprivation, and xylem pattern development (Schoppach et al. 2016). While there are other, less strong QTLs that control SLP in the RAC875 × Kukri population, those findings are consistent with a hypothesis of a root-based hydraulic restriction underlying the water saving behavior of RAC875.

Another key finding of the genetic analysis of SLP and other traits examined in Schoppach et al. (2016) was that maturity genes Ppd-B1 and Ppd-D1 significantly influenced all of the examined traits including SLP, despite the narrow flowering range of the population of 143 DH lines. In addition, compared to Ppd-B1, the effect of Ppd-D1 was much stronger on SLP, indicating different mechanisms through which development could influence drought tolerance traits. Overall, those findings delineate a relatively low number of QTL controlling SLP which is a favorable outcome in terms of pyramiding favorable alleles in a breeding program.

References

Bennett D, Reynolds M, Mullan D, Izanloo A, Kuchel H, Langridge P, Schnurbusch T (2012) Detection of two major grain yield QTL in bread wheat (*Triticum aestivum* L.) under heat, drought and high yield potential environments. Theor Appl Genet 125:1473–1485

Bonneau J, Taylor J, Parent B, Bennett D, Reynolds M, Feuillet C, Langridge P, Mather D (2013) Multi-environment analysis and improved mapping of a yield-related QTL on chromosome 3B of wheat. Theor Appl Genet 126:747–761

Bramley H, Turner NC, Turner DW, Tyerman SD (2009) Roles of morphology, anatomy, and aquaporins in determining contrasting hydraulic behavior of roots. Plant Physiol 150:348–364

Condon AG, Farquhar GD, Richards RA (1990) Genotypic variation in carbon isotope discrimination and transpiration efficiency in wheat. Leaf gas- exchange and whole-plant studies. Aust J Plant Physiol 17:9–22

Condon AG, Richards RA, Rebetzke GR, Farquhar GD (2002) Improving intrinsic water-use efficiency and crop yield. Crop Sci 42:122–131

Curtis BC (2002) Wheat in the world. In: Curtis BC, Rajaram S, Gomez Macpherson H (eds) Bread wheat: improvement and production. Food and Agriculture Organization of the United Nations, Rome

Engels C, Fuganti-Pagliarini R, Marin SRR, Yamaguchi-Shinozaki K, Nepomuceno AL (2013) Introduction of the rd29A:AtDREB2A ca gene into soybean (*Glycine max* L. Merril) and its molecular characterization in leaves and roots during dehydration. Genet Mol Biol 36:556–565

FAO (2011) Crop prospects and food situation. Food and Agriculture Organization, Global Information and Early Warning System, Trade and Markets Division (EST), Rome

FAO (2016) FAO Cereal supply and demand brief. http://www.fao.org/worldfoodsituation/csdb/en/. Accessed 23 Dec 2016

Farooq M, Hussain M, Siddique KHM (2014) Drought stress in wheat during flowering and grain-filling periods. Crit Rev Plant Sci 33(4):331–349

Farquhar GD, Richards RA (1984) Isotopic composition of plant carbon correlates with water-use efficiency of wheat genotypes. Aust J Plant Physiol 11:539–552

Fischer RA, Rees D, Sayre KD, Lu Z, Condon AG, Saavendra AL (1998) Wheat yield progress is associated with higher stomatal conductance, higher photosynthetic rate and cooler canopies. Crop Sci 38:1467–1475

Fleury D, Jefferies S, Kuchel H, Langridge P (2010) Genetic and genomic tools to improve drought tolerance in wheat. J Exp Bot 61:3211–3222

Izanloo A, Condon AG, Langridge P, Tester M, Schnurbusch T (2008) Different mechanisms of adaptation to cyclic water stress in two south Australian bread wheat cultivars. J Exp Bot 59:3327–3346

Kholová J, Nepolean T, Hash CT, Supriva S, Senthilvel S, Kakkera A, Yadav R, Vadez V (2012) Water saving traits co-map with a major terminal drought tolerance quantitative trait loci in pearl millet [*Pennisetum glaucum* (L.) R. Br.] Mol Breed 30:1337–1353

Kirkegaard JA, Hunt JR (2010) Increasing productivity by matching farming system management and genotype in water-limited environments. J Exp Bot 61:4129–4143

Kudoyarova G, Veselova S, Hartung W, Farhutdinov R, Veselov D, Sharipova G (2011) Involvement of root ABA and hydraulic conductivity in the control of water relations in wheat plants exposed to increased evaporative demand. Planta 233:87–94

Mahdieh M, Mostajeran A (2009) Abscisic acid regulates root hydraulic conductance via aquaporin expression modulation in *Nicotiana tabacum*. J Plant Physiol 166:1993–2003

Nouri G (2013) The making of the Tunisian revolution: contexts, architects, prospects. Oxford University Press, UK. ISBN 978-0-7486-9103-6

Rebetzke GJ, Richards RA (1999) Genetic improvement of early vigour in wheat. Aust J Agric Res 50:291–301

Rebetzke GJ, Condon AG, Richards RA, Farquhar GD (2003) Genetic control of leaf conductance in three wheat crosses. Aust J of Agr Res 54:381–387

Rebetzke GJ, Rattey AR, Farquhar GD, Richards RA, Condon AG (2013) Genomic regions for canopy temperature and their genetic association with stomatal conductance and grain yield in wheat. Funct Plant Biol 40:14–33

Richards RA, Passioura JB (1989) A breeding program to reduce the diameter of the major xylem vessel in the seminal roots of wheat and its effect on grain yield in rain-fed environments. Aust J Agric Res 40:943–950

Rosegrant M, Cai X, Cline S, Nakagawa N (2002) The role of rainfed agriculture in the future of global food production. International Food Policy Research Institute, Discussion Paper NO. 90.

Schoppach R, Sadok W (2012) Differential sensitivities of transpiration to evaporative demand and soil water deficit among wheat elite cultivars indicate different strategies for drought tolerance. Environ Exp Bot 84:1–10

Schoppach R, Wauthelet D, Jeanguenin L, Sadok W (2014) Conservative water use under high evaporative demand associated with smaller root metaxylem and limited trans-membrane water transport in wheat. Funct Plant Biol 41:257–269

Schoppach R, Taylor JD, Majerus E, Claverie E, Baumann U, Suchecki R, Fleury D, Sadok W (2016) High resolution mapping of traits related to whole-plant transpiration under increasing evaporative demand in wheat. J Exp Bot 67:2847–2860

Schoppach R, Fleury D, Sinclair TR, Sadok W (2017) Transpiration sensitivity to evaporative demand across 120 years of breeding of Australian wheat cultivars. J Agron Crop Sci doi:10.1111/jac.12193 (in press)

Walton JK, Seddon D (2008) Free markets and food riots: the politics of global adjustment. Wiley, Chichester/New York. ISBN 978-0-470-71271-9

Wasson AP, Richards RA, Chatrath R, Misra SC, Sai Prasad SV, Rebetzke GJ, Kirkegaard JA, Christopher J, Watt M (2012) Traits and selection strategies to improve root systems and water uptake in water-limited wheat crops. J Exp Bot 63:3485–3498

Index

A
Abscisic acid (ABA), 78
APSIM model, 66
Aquaporins (AQP)
  de-rooted soybean shoots, 21
  hydraulic conductivity, 30–31, 41
  pearl millet (*Pennisetum glaucum*), 77
  PI 416937, 20, 21
  silver treatment, 21
  sorghum (*Sorghum bicolor*), 70
  transpiration rate, 21
  VPD, 22
  wheat (*Triticum aestivum*), 88, 89

C
Cell-to-cell pathway, 77
Central metaxylem diameter (CMX), 89
Chickpea (*Cicer arietinum*)
  AQP, 41, 42
  drought, 35
  hydraulic limitation, 39, 40
  transpiration response
    soil drying, 36–37
    VPD, 37–39
  water extraction, 42
  water transport pathways, 41, 42
Crop yield, 9, 60

D
Drought
  cycle, 2
  maize, 55
  peanut, 27
  sorghum growth, 66
  tolerance, 47, 75
Drought stress index (DTI), 43

F
Focused identification of germplasm strategy (FIGS), 50
Fraction of transpirable soil water (FTSW), 6, 7, 9, 27, 48, 55, 56, 86

G
Gene expression, 78–79
Genetic variation, 13, 14, 22, 59
Germplasm release, 23

H
Hydraulic conductivity, 6, 7, 15, 18, 30–31, 40, 59, 88, 89

L
Lentil (*Lens culinaris*)
  drought stress, 47
  FTSW, 49
  water conservation
    soil drying, 48, 49
    VPD, 49, 50
  water-saving traits, 50, 52, 53
Limited transpiration
  benefits, 12, 13
  crop species, 15
  genetic variation, 13, 14, 22

T.R. Sinclair (ed.), *Water-Conservation Traits to Increase Crop Yields in Water-deficit Environments*, SpringerBriefs in Environmental Science,
DOI 10.1007/978-3-319-56321-3

Limited transpiration (*cont.*)
genotypes, 15
low hydraulic conductivity, 15
and normal transpiration rate, 12
TR, 11
VPD, 11, 12, 57, 66–69
Linear-segment regression model, 39

M

Maize (*Zea mays*)
aquaporin inhibitors, 59
crop yield, 60
drought, 55
hydraulic conductivity, 59
transpiration
high temperatures, 57
soil drying, 55, 56
TR_{lim} trait, 57
VPD, 56–58
water deficit, 60

N

Normalized transpiration ratio (NTR), 48, 56

P

Peanut (*Arachis hypogaea*)
TR_{lim} trait, 31–33
water conservation
soil drying, 27, 28
VDP (*see* Vapor pressure deficit (VDP))
water deficits, 27
Pearl millet (*Pennisetum glaucum*)
abscisic acid (ABA), 78
gene expression, 78–79
water conservation
crop yield, 81
soil drying, 74
VPD, 75
water transport pathways
apoplastic pathway, 77
AQP, 77, 78
hydraulic conductivity, 78
symplast pathway, 77
TR response, 76–78
transcellular pathway, 77

R

Recombinant inbred lines (RIL), 22

S

Simple simulation model (SSM), 8, 36
Slow-wilting phenotype, 19
Soil drying
peanut, 28
stomatal closure
FTSW, 9
soil drying, 5, 7
soybean, 8
SSM, 8
transpiration response
FTSW, 37
SSM, 36
VPD, 37–39
water deficit, 36, 37
water conservation, 17, 18
Sorghum (*Sorghum bicolor*)
AQPs, 70
hydraulic conductivity, 69
soil drying
grain and stover yield, 66
sorghum growth, 66
transpiration response, 66
VPD
environmental conditions, 69
limited-transpiration rate, 66–69
simulation models, 68
transpiration rate (TR), 67–69
Soybean (*Glycine max*)
advantage, 17
aquaporins (AQP), 20–22
germplasm release, 23
RIL, 22
water conservation
soil drying, 17, 18
vapor pressure deficit, 19–20
yield advantage, 23–25
SSM. *See* Simple simulation model (SSM)
Stomatal closure
drought, 2
FTSW, 7, 9
soil drying, 5, 7
soybean, 8
SSM, 8
transpiration rate, 7
Stomatal conductance, 1, 7, 66, 67, 69, 78

T

Transpiration
FTSW, 6
lentil (*Lens culinaris*), 48
soil drying, 55, 56

VPD, 56–58
water-limited conditions, 1
Transpiration rate (TR), 11, 74
TR_{lim} trait, 31, 33

V
Vapor pressure deficit (VPD)
aquaporins, 30–31
environmental conditions, 31
hydraulic conductivity, 30–31
lentil (*Lens culinaris*), 49, 50
limited-transpiration trait, 12
vs. transpiration rate, 32
transpiration response
plant growth rate, 39
stomatal conductance, 39
water-deficit-sensitive genotypes, 37, 38
TR_{lim} trait, 29, 30
water-limitation trait, 28–30

W
Water conservation
crop yield, 81
lysimetric system, 42
peanut, 28
soil drying, 17–18
lentil (*Lens culinaris*), 48–49
pearl millet (*Pennisetum glaucum*), 74–75
wheat (*Triticum aestivum*), 86
stomatal closure, 2
VPD
genotype response, 90
pearl millet (*Pennisetum glaucum*), 75–76
slow-wilting phenotype, 19
SLP, 91
TR response, 90–91
water-limitation trait, 19–20
wheat (*Triticum aestivum*), 86, 87
Water-deficit stress, 5, 36, 37, 42
Water-limitation trait, 1, 2, 19, 20
Water-saving traits, 50, 52, 53, 78–79
Water transport pathways
apoplastic pathway, 77
AQP, 77, 78
hydraulic conductivity, 78
symplast pathway, 77
TR response, 76–78
transcellular pathway, 77
Wheat (*Triticum aestivum*)
AQPs, 88, 89
drought, 85
leaf-based hydraulic restriction, 88
water conservation
soil drying, 86
VPD, 86, 87

FSC
www.fsc.org
MIX
Papier | Fördert gute Waldnutzung
FSC® C083411

Kolibri 360 GmbH
Ferdinand-Jühlke-Straße 7
99095 Erfurt, Deutschland
produktsicherheit@kolibri360.de